CW00351946

Agile Business Analysis

Enabling Continuous Improvement
of Requirements, Project Scope,
and Agile Project Results

Kevin Aguanno
PMP, PMI-ACP, CSM

Ori Schibi
PMP, PMI-PBA, PMI-ACP, SMC

Library of Congress Cataloging-in-Publication Data
Names: Aguanno, Kevin, author. | Schibi, Ori, 1971- author.
Title: Agile business analysis : enabling continuous improvement of
    requirements, project scope, and agile project results / by Kevin Aguanno
    and Ori Schibi.
Description: Plantation, FL : J. Ross Publishing, [2018] | Includes index.
Identifiers: LCCN 2018025763 (print) | LCCN 2018034200 (ebook) | ISBN
    9781604278019 (e-book) | ISBN 9781604271485 (hardcover : alk. paper)
Subjects: LCSH: Business analysts. | Business planning. | Agile software
    development.
Classification: LCC HD69.B87 (ebook) | LCC HD69.B87 .A48 2018 (print) | DDC
    658.4/012—dc23
LC record available at https://lccn.loc.gov/2018025763

Phone: (954) 727-9333
Fax: (561) 892-0700
Web: www.jrosspub.com

To all of those who have struggled through their own agile transformations:

May your efforts inspire others to implement changes in order to ensure that those who follow you will find that their paths forward have been smoothed.

CONTENTS

FOREWORD

From the moment an organization begins to adopt agile strategies, everyone asks a fundamental question: "How do I fit in?" This is an easy question to answer if you're currently a developer or tester, but it may be a little harder if you're a designer or architect, and could be rather difficult if you're a manager or business analyst.

It's completely natural to ask how you fit into agile and, frankly, I'd be concerned if you didn't ask that question. When I'm working with people who are trying to become agile, I recommend that they begin by exploring these questions: "How is agile different?" and "What is the agile mindset?" The bad news is that it will take you months, and perhaps even years, to truly internalize the answers to these questions. The good news is that you can easily start learning how to *be agile* if you're open to new (and proven) agile ideas. This new mindset includes a desire for close collaboration with both stakeholders and teammates; a desire to reduce the feedback cycle from the time that you hear an idea to the time it's implemented; a desire to reduce waste by focusing on effective communication and executable specifications over static specifications; and a desire to broaden your skill set beyond business analysis and to share your valuable analysis skills with others. Hopefully, you are doing some of these things already.

It isn't sufficient to just *be* agile, regardless of what some agile coaches may claim—you need to know how to *do* agile, too. This leads us to critical questions such as: "How does business analysis fit into agile?" "Who is doing this work?" and "What skills do I need to add value on agile teams?" This is where things get interesting for traditional business analysts. In order, the initial answers to these questions are: analysis is so important on agile teams that we do it throughout the entire life cycle, although we might not need to do it every single day; the person(s) with the skills and ability (who may or may not be an agile business analyst); and lightweight agile modeling and documentation skills, to start.

This leads us to the crucial question: "When would a business analyst be needed on an agile team?" The honest answer is: "Not always"—although perhaps a more palatable answer is: "Whenever the situation is sufficiently complex to warrant a specialized person in that role."

My experience, after working with dozens of agile teams over almost two decades on five continents, is that agile business analysts are needed on agile teams that are working at scale. This includes agile teams that are taking on complex domain problems, large teams (often organized as teams of teams), teams in life-critical or regulatory environments, and on geographically distributed teams. Agile methods such as Scrum don't address such scaling challenges well and don't include the role of business analyst as a result or, perhaps because Scrum doesn't include a business analyst role, it doesn't scale well in practice— you be the judge. Interestingly, the Large-Scale Scrum (LeSS) method, which is geared for teams of teams that are often geographically distributed, does include an explicit role of business analyst. The Disciplined Agile (DA) framework also suggests having business analyst specialists on teams when you run into any of the previously mentioned complexities.

So what's my point? The authors of this book answer these questions, and more, in detail. I believe that *Agile Business Analysis: Enabling Continuous Improvement of Requirements, Project Scope, and Agile Project Results* will prove to be a valuable resource on your learning journey. Read it now and take its advice seriously.

Scott Ambler
April 2018

Scott Ambler is an enterprise transformation coach and senior consulting partner at Scott Ambler + Associates (ScottAmbler.com). Scott led the development of the agile modeling (AM) (agilemodeling.com) and agile data (agiledata.org) methods in the early 2000s and is a co-creator of the DA framework (disciplinedagiledelivery.com).

INTRODUCTION

The decision to write this book came as a result of both of the authors' experience in agile project environments and their observations about the emerging trends and challenges in project management, business analysis, and the application of agile concepts. Kevin Aguanno's experience puts him at the forefront of adapting agile methodologies since he has been involved with these practices since before they were called agile. He was the first person to coin the term *agile project management*, and he has since helped shape the agile world by pioneering the concepts of disciplined agile delivery. Ori Schibi's experience includes introducing agile concepts in multiple settings and pragmatically adapting the approach to the specific needs of projects and the individual organizations. Ori is also a thought leader in the area of improving the collaboration between the project manager (PM) and the business analyst (BA).

Both authors' expertise extends beyond the application of agile concepts, and includes performing agile organizational readiness assessments that identify areas to improve and adjust in order to allow agile to succeed. This helps organizations choose the right approach (agile versus waterfall or something in between) for projects, and incorporating business analysis to support the successful delivery of agile projects.

Through their experiences, both authors noticed a growing tension in the background of agile projects; a struggle that reflects two growing and, at times, conflicting trends that seem to clash in the minds of many: agile and business analysis. Both agile and business analysis have grown in parallel, but the application of these areas takes place, for the most part, in different settings. Many organizations utilize BAs for both project and non-project work. The formation of the International Institute of Business Analysis (IIBA) in 2003 introduced a new era for the business analysis profession as it was the first attempt to define the role of the BA—with an enhanced focus on requirements. The Project Management Institute (PMI) followed through and introduced the Collect Requirements Process as part of Project Scope Management before further expanding the role of business analysis with the introduction of the PMI-PBA

Certification. The increased focus on business analysis further clarified the role of the BA and his/her involvement within projects and at the enterprise level.

With the rise of the BA role came obvious challenges, such as the blurry delineation of responsibilities between the PM and the BA. The lack of specific role definitions has led to friction, misunderstandings, gaps, and duplication of efforts as both worlds were colliding.

While the growth in business analysis was taking place, agile started to pick up, but in a parallel trajectory. Agile does not call specifically for BAs to be included in the projects and, in fact, the entire setup of agile projects appears to be designed to eliminate the need for intermediaries such as the BA. In the early days (after the introduction of the term agile and the *Agile Manifesto* in 2001) it appeared that there was no need for BAs. Early adaptations of agile were mostly confined to technology projects that were small. Team members were handpicked and the level of scrutiny was high. Over the years, the use of agile started to expand rapidly and become widely accepted. Agile has been scaling and growing—from small to all project sizes; from technology projects to pretty much any industry; and from more nimble, risk-taking, small, and change-hungry organizations to more traditional settings such as large financial institutions, governments, and other more conservative and slow-moving organizations.

It is no secret that even today many agile practitioners believe that there is no room or need for a BA on agile projects. On paper they may be right; however, there are two distinct drivers that point at agile projects' ability to benefit from BAs, or at least from the application of business analysis skills:

1. There are many areas in agile projects that have built-in challenges. These challenges can be covered and looked after with the application of business analysis skills. Many organizations struggle with getting clear mandates from their product owners; ownership issues around backlog maintenance; coordination between the business, testers, and developers; task updates; work allocation; and escalation. In well-functioning agile teams, nearly all team members should demonstrate basic business analysis skills; in reality, however, this may not be the case. Many teams struggle with the coordination of these issues and the true ownership of handoffs and other tasks that *fall* under the realm of business analysis.

2. The rapid and widespread adoption of agile has introduced new challenges that can benefit from business analysis skills. In the past, if agile team members were *handpicked* to ensure their skills and experience levels were applicable to the agile project's needs, the growth of agile ensures it becomes more difficult to guarantee that all team members are suitable for this kind of environment. Throughout this book, we establish that agile is a very unforgiving approach. There is a need for all

team members and stakeholders to *play along* and perform their roles, and for processes, policies, and governance structures to be aligned with an agile project's needs. When people or processes around agile projects are not aligned with the agile project, things will unravel quickly, causing project performance to lag. With the fast pace and the challenges that agile methods introduce, there is a consistently growing need for a BA to provide another set of eyes looking for gaps and picking up the pieces when things are misaligned. Although it is expected that business analysis activities are shared by all team members, in many agile environments, team members do not perform business analysis skills consistently, or even at all. Further, even when team members demonstrate business analysis activities, there are two types of challenges that surface: (1) when team members wear two hats (for example when their roles are split between being a developer and a BA), the BA role is secondary and when there are issues or time constraints, the secondary role typically gets pushed aside and (2) when multiple team members are responsible for business analysis activities, the seams and touchpoints between team members tend to become rough and inconsistent.

BOOK OBJECTIVES

This book examines the benefits that agile projects can gain by ensuring that there are business analysis skills present and that team members demonstrate them consistently. The book does not specifically call for a dedicated BA role in projects; however, it identifies many areas where a BA can complement the skills and experience of other team members and benefit the project. In fact, BAs can help support the product owner, the Scrum Master (or PM), and individual team members (ranging through all roles).

This book introduces considerations and criteria to determine whether it is sufficient to grow business analysis skills among team members or whether it is necessary to have an individual who has a BA title. The intent of the book is to advocate that whatever form is used, projects can benefit from business analysis skills. The book promotes the need to focus on bringing these two growing areas together—agile and business analysis—not by force but rather as a growing necessity.

Pride, ego, and misconceptions can often cloud people's judgment, and even the most knowledgeable and experienced individuals often get locked in paradigms and preconceived notions that prevent them from realizing opportunities. Many agile practitioners are adamant in their refusal to introduce BAs into agile projects, but an increasing number of practitioners realize the need for business analysis skills within their teams.

We believe that the biggest barrier is ensuring that there is recognition of the need for business analysis skills and, once acknowledged, there can be a healthy and productive discussion on how to best introduce these skills into agile projects. This book facilitates this discussion and helps decision makers reach the right conclusion based on the needs of the project, the organization, and the team. Such discussions, if done properly, incorporate all surrounding conditions, circumstances, constraints, and issues that agile teams—and managers—face.

WHAT MAKES THIS BOOK DIFFERENT

This book reviews agile concepts and looks at how organizations can benefit from agile delivery approaches, if applicable to their needs. The considerations on whether to use agile are viewed from a business analysis perspective. The book will help you to decide *if* and *where* business analysis concepts can be applied to help agile projects, and explores criteria determining whether the business analysis skills should be applied throughout the team or by introducing a BA to be part of the team.

The authors bring their unique styles and experiences with applying disciplined approaches to agile into the discussions around the need for business analysis skills. Their pragmatic approach does not try to impose ideas on the reader; rather, it focuses on enabling the reader to make the right decision as if the book is tailored to the readers' needs and circumstances.

THE STRUCTURE OF THE BOOK

Chapter 1 provides an introduction to agile and agile processes. It provides basic comparisons between waterfall and agile, continues with a review of agile concepts, and supports the review with a discussion of lean principles. Throughout the chapter, several ideas on how to minimize waste with the help of business analysis are introduced.

Chapter 2 discusses agile challenges. Agile provides many answers to the challenges introduced by Waterfall approaches, but agile is not perfect. First, it is important to recognize that agile has its own challenges; then it is time to review these challenges and learn how not to stumble on them—and if we do stumble, how to overcome these challenges. The primary challenges that this chapter covers include different views of what is agile, misconceptions and challenges related to culture, and organizational change. After a review of the *Agile Manifesto*, the chapter discusses how to deal with the organizational change elements as a result of agile adaptations, providing an approach for agile readiness and governance, and ensuring that the organization chooses the right approach. The chapter closes by reviewing additional challenges (and ways to overcome them) including focusing too much on the process, resource issues in a matrixed organization, the *diluting effect* that agile projects face due to rapid growth, and finally biases and challenges related to the Scrum method.

Chapter 3 examines roles and responsibilities in agile environments and shows the potential of the BA to help overcome these challenges. The chapter then moves to a more specific definition of the role of the BA in agile environments, what to expect when there is a BA assigned to agile projects, and role impact: how the BA supports and impacts the traditional agile roles of product owner, Scrum Master, and team members.

Chapter 4 reviews agile requirements processes, including the use of user stories. After an overview of the topic, the chapter takes the readers through a discussion of the relationship between user stories and use cases and how they support the agile requirements process. Beyond exploring use cases (both scenarios and diagrams) and the role of the BA in supporting their creation and maximizing the benefits they produce, the chapter reiterates the differences between *traditional* and agile requirements and shows how the BA can provide valuable support for the product owner. The book reviews different types of requirements to ensure no requirement *type* is overlooked and examines other requirements-related elements including themes, epics, and modeling.

Chapter 5 deals with agile documentation, a topic that is greatly misunderstood. Many people still believe *agile* means creating no documentation (in addition to no planning) and working fast, but these are all misconceptions. The chapter illustrates the importance of documentation in agile projects, shows how the BA plays an instrumental role in ensuring that the right documentation is created at the right level of detail, and looks into agile reporting, artifacts, project reports, and information radiators.

Chapter 6 focuses on the role of the BA in planning and estimating, which is different than the role a BA would fulfill in traditional waterfall projects. The BA supports the estimating and planning process by demonstrating the same skills he or she would in waterfall projects; however, the application of the skills in agile projects is different. This chapter also reviews business analysis activities that try to enhance the focus of the product owner and then analyzes the different planning cycles in agile projects and how they can benefit from a good business analysis—release planning, iteration planning, and daily planning. Other important concepts are also discussed through the lenses of how they can benefit from a BA: the prioritization process and techniques to estimate complexity and team velocity.

Chapter 7 provides an additional review of the prioritization process and looks for ways to improve processes around backlog maintenance, end of iteration demos, retrospectives, and applying lessons learned to upcoming iterations.

Chapter 8 covers the area of agile testing and solution evaluation. With testing often becoming an issue in agile projects due to the increase in the importance, timing, involvement, and scope of the testing, it becomes an area that can benefit significantly from the support of a BA. The chapter opens with a review of the agile approach to quality and it proceeds through a discussion of the expanded role of testing in agile projects. Additional areas this chapter covers include the all-important *definition of done*, the BA's support of the testing process, how to deal with defects, and challenges associated with agile testing. Due to the increased amount of testing on agile projects and the need for regression testing in every iteration, this chapter also looks into testing automation considerations and the benefits a BA can add in this area. The chapter concludes with a look into agile testing best practices and agile testing strategies.

Chapter 9 is one of the most unique chapters as it breaks down the day of the agile BA—item by item—as he or she supports the agile team. The daily breakdown does not attempt to dictate to the reader a prescription; rather, the chapter identifies typical activities and how they may all come together on a sample project. Your own project may require a different configuration; however, the sample provided illustrates the concepts to help you identify how to integrate agile business analysis practices on your own project. The chapter identifies three template days in the life of an agile BA: *the last day of an iteration*—where the focus is on wrapping up the iteration, demonstrating the product, getting feedback, incorporating feedback into the reprioritization process, and learning from what just happened; *the first day of a new iteration*—where the BA supports the team finalizing the planning and the setup of the new iteration based on the feedback and the results of the previous iteration; and *a typical middle day in the iteration.*

Chapter 10 cycles back to agile challenges and business analysis, but this time views them through a more discerning lens. The chapter discusses agile and social trends, reviews concepts around organizational agility, and looks into the future of agile and project management, the future of business analysis in agile projects, and the prospect of turning the BA into an agile center of excellence. The chapter concludes by reminding the readers that a BA must not act as a replacement for the product owner.

HOW TO BENEFIT FROM THIS BOOK

This book is an easy read and readers can focus on one chapter at a time. It has many practical examples and provides multiple practical ways to apply concepts, translate ideas and theories into benefits, and improvements to the way we manage agile projects. The content of this book reflects the authors' education and years of experience with applying the information, concepts, and ideas that are presented here—with proven results.

ACKNOWLEDGMENTS

Kevin Aguanno

My own journey to find more flexible and efficient ways of delivering high-change projects began in the 1980s. Early inspirations included Barry Boehm, James Martin, and Steve McConnell. Later, the writings of Kent Beck, Jeff Sutherland, and others provided additional insights. These (and others) led me to experiment with different approaches to defining and developing software-based systems and to combine these approaches with the project management techniques needed to ensure compliance with organizational governance processes while still allowing the project teams to be efficient in how they adapt to change. Despite my own pioneering work of combining agile estimating, planning, monitoring, and control techniques with traditional project management practices, I cannot take credit from where it is really due—I stand on the shoulders of giants in the field who set the groundwork for my own innovations.

I would also like to thank my co-author, Ori Schibi. It was his long hours and dedicated work that really made this book project happen.

Last, but not least, I would like to thank my wife, Alba, and my two children for their patience, support, and love.

Ori Schibi

To my co-author, Kevin Aguanno—a friend, a colleague, and an inspiration: it is a true pleasure knowing you and working with you and for you.

To my wonderful wife, Eva (who puts up with me), and our delightful daughters, Kayla and Maya: your support, the happiness you bring, and your love are more than what anyone could ask for.

To my loving mom, Hava, and brothers, Eitan and Naaman.

In Loving Memory of Ruti Dalin (1939–2018), whom I have known my entire life: you were the closest thing to family there is. We will miss you and love you always.

ABOUT THE AUTHORS

Kevin Aguanno, PMP, PMI-ACP, CSM, has over 20 years of experience managing complex systems integration and software development projects. He is well known in the industry for his innovative approaches to solving common project management problems. He focuses on two project management specialty areas: agile project management and troubled project recovery.

As a well-known keynote speaker, author, trainer, and coach in agile management methods, Aguanno has taught thousands of people how to better manage high-change projects by using techniques from Scrum, Extreme Programming, Feature-Driven Development, and other agile methods. He is a frequent presenter at conferences and private corporate events where he delights audiences with practical advice peppered with fascinating stories from his own experiences in the trenches practicing agile project management. He is also founding Director and Vice President of the Project Management Association of Canada. He resides in Toronto, Canada.

Ori Schibi is President of PM Konnectors, an international consulting practice based in Toronto. With focus on project management, business analysis, and change management, his company provides a variety of services, including facilitation, workshops, training, and consulting—all of which deliver value to clients through a wide range of innovative business solutions. Large to mid-sized organizations in diverse industries and all levels of governments have benefited from PM Konnectors innovative services.

Schibi, PMI-PBA, PMP, PMI-ACP, SMC, is a thought-leader and subject matter expert in business analysis and both traditional and agile project management. He is a published author, speaker, and consultant with over 25 years of proven experience in driving operational and process improvements, software implementations, and complex programs to stabilize business, create growth and value, and lead sustainable change. He resides in Ontario, Canada.

Web
Added
Value™

This book has free material available for download from the
Web Added Value™ resource center at *www.jrosspub.com*

At J. Ross Publishing we are committed to providing today's professional with practical, hands-on tools that enhance the learning experience and give readers an opportunity to apply what they have learned. That is why we offer free ancillary materials available for download on this book and all participating Web Added Value™ publications. These online resources may include interactive versions of material that appears in the book or supplemental templates, worksheets, models, plans, case studies, proposals, spreadsheets and assessment tools, among other things. Whenever you see the WAV™ symbol in any of our publications, it means bonus materials accompany the book and are available from the Web Added Value Download Resource Center at www.jrosspub.com.

Downloads for *Agile Business Analysis: Enabling Continuous Improvement of Requirements, Project Scope, and Agile Project Results* consist of:

- An Agile Triage Checklist
- An Example of an Iteration Burndown Spreadsheet
- An Agile Glossary
- PowerPoint Slides that Cover Key Agile Business Analysis Concepts
- Educational White Papers on:
  - Agile DevOps
  - Agile Methods and the Need for Speed
  - Agile and Capital Cost Appreciation

1

INTRODUCTION TO AGILE
AND AGILE PROCESSES

Agile is not new, but its growth and increasing rate of adoption introduce new challenges (and opportunities) on an ongoing basis. Agile is also not the solution for all problems, but it is an approach, an umbrella, and a state of mind that, if done properly, can yield significant benefits for the performing organization. With that said, if agile is introduced the wrong way—in an unsuitable environment or under the wrong circumstances—it will cause more harm than good and there will always be someone who puts the blame on agile. While Chapter 2 of this book deals with challenges that emerge as part of adopting agile methods, this chapter introduces key agile concepts and examines the role of business analysis in delivering agile project success.

ABOUT AGILE AND WATERFALL

Agile is a pragmatic approach that recognizes that both the project team and the client do not know everything that has to be done up front, and that things will change along the way. Agile is an empirical approach—or, in other words, observation-based. It is open for changing needs and is evolutionary by nature, with multiple iterations (or sprints)—each producing a *shippable* product increment. The iterations are grouped into releases, where the customer actually receives, accepts, and, if needed, uses the released product increment. Agile addresses many of the challenges that were introduced by practices related to the waterfall approach, also known as a deterministic approach, where the team finalizes the requirements early on and then proceeds to build the agreed-upon product. Agile is made of the combination of two approaches: incremental (where product increments are released along the way) and iterative (where adaptation to change is continuous). Agile is not simply about breaking a large project into many small waterfall phases; there is more to it, thanks to the combination between the incremental and the iterative approaches.

It is not that agile is, plain and simple, better than waterfall. Although it addresses many bad practices that have been enabled by waterfall, there is still a place and time for a more deterministic approach. If a project has a set scope that is not about to change, and if the client does not need to benefit from early releases of the product by using it along the way, a waterfall approach may be suitable. Also, the fact that agile tries to fix some of the problems associated with waterfall does not mean that waterfall is at fault for project problems and is to blame for project failures. Waterfall simply enables some bad behaviors by allowing them to *hide* for too long before there are checks and balances to realize them. For example, because there are no interim releases and iterations along the way, the customer has the ability to see the product only at the very end— after testing has been completed—and only then can feedback be offered. It is possible that along the way the reports that provided the customer with updates were not portraying the actual picture, but the client had no chance to realize it because the reports were the only thing that the client had as *proof* of progress. This does not imply that the project team deliberately misled the client, but it is possible that information was communicated in a way that was misunderstood by the client. Good business analysis work and effective communication between the project team and the client can minimize the risk of such problems, but it provides no guarantee against such communication gaps.

Agile is an approach with several methodologies under its *umbrella*. Some of these methodologies include Scrum, Feature Driven Development, Extreme Programming (XP), Dynamic Systems Development Method (DSDM), Disciplined Agile Development, and Lean Development. While deterministic approaches focus on the plan and provide less flexibility, they also provide less chance of meeting schedule and cost constraints, which might lead to time and cost pressures that in turn may lead to quality problems. Empirical approaches, on the other hand, are evolutionary, observation based, and focus on delivering value by *locking* time, cost, and quality goals and focusing on scope flexibility.

In the early days of agile, around 2001, when agile was named *agile*, there was little to no talk about business analysis, or business analysts (BAs) in agile environments—and it was for good reasons:

- The role of the BA was still undefined—it was prior to when the International Institute of Business Analysis was founded (in 2003) and prior to when the Project Management Institute expanded its focus to include business analysis.

- The main premises of agile—including its focus on handling change, being nimble, reducing waste, improving communication, and building teams that are more cohesive—are all about reducing the need for a BA to *bridge* and connect different stakeholders, groups, and teams. It was therefore expected that agile role definitions include only three roles: (1) the product owner; (2) the Scrum Master, coach, or a role that is somewhat equivalent to that of a project manager (PM); and (3) a team member (meaning all technical members of the team). While some interpretations call for the inclusion of a BA as part of the team, this was not the original intent. Another interpretation of agile concepts agrees that there is a need for business analysis skills, but this need has to be addressed by team members demonstrating business analysis skills, without calling for a dedicated or distinct role of a BA on the project.

This book covers the growing and increasingly recognized need to have the skills of the BA present on agile projects. At times, we will discuss the need for business analysis skills to be demonstrated by a BA, and at other times, ensure that team members can demonstrate these skills. One way or another, it is clear that these skills are necessary and are important for project success, but there is no one clear and correct way to apply them. This means that for some project environments, it would be sufficient to have team members demonstrating business analysis skills, while other project environments need a BA to champion these skills and ensure they are applied consistently and correctly. In retrospect, as agile is closing in on completing its second decade, it is safe to say that although agile attempts to produce benefits that, if performed properly, lead to a reduction, if not total elimination of the need of BAs, agile growth and the way it is applied bring back those needs and with them—at times—the need for a specific role of a BA within the team.

One of the reasons that has rejuvenated the focus on business analysis skills in agile environments is the rapid growth of agile, its scalability beyond technology projects, and the attempts to introduce agile concepts in organizations and environments that are not a natural fit for agile—including slower moving financial institutions and government agencies. With more projects attempting to apply agile concepts and more team members assigned to those projects, there is a need to refocus the efforts and to add checks and balances that reduce the chance that these projects will fail. When more people within the organization practice agile and when more agile projects are competing for scarce resources, it leads to what we identify as *agile dilution*, where there is a decline in teams' relevant agile skills and a reduction in the true agile experts in the organization to oversee, support, and lead the agile process.

Before proceeding to discuss the basic concepts of agile, it is important to make the record clear: agile is *not* about the following things:

- Not about *working fast*: but rather, building products in such a way that we can produce value earlier.
- Not about *no planning*: agile planning is critical for success and it is simply done differently and throughout the project, instead of a heavy focus on building a plan early on. Agile focuses on the act of planning, rather than on the creation of a plan.
- Not about *no requirements*: similar to the planning, an important part of agile is requirements. While they are referred to by a different name—user stories—there is a need to elicit and manage requirements. It is done in a different fashion than what takes place in waterfall environments, but properly managing requirements is a critical part of agile success. It is the level of detail around requirements that changes since the project starts with high-level requirements and then the team elaborates and seeks details on a timely basis—like a rolling wave, only for the requirements that are up next in the cycle. The team and stakeholders need to come to terms with the fact that ambiguity is welcomed until more information is needed. In a way, the requirements are partially trial and error because the requirements that are defined at the start of the project may need to change later on as more information is discovered during the detailed planning. Agile is no different than waterfall when it comes to ambiguity. In both types of approaches, there is ambiguity and there is no full set of detailed information up front—it is just that agile teams recognize the ambiguity and accept it, while in waterfall, there is an attempt to finalize requirements and their details up front. This latter effort costs a lot of time, effort, and money, only to go through the likely need to change these requirements later on—in a process that not only costs a lot, but also one that adds a lot of risk. The reference made here to trial and error does not refer to the project objectives or the project vision, but rather to the process of building stories and refining their details.
- Not about *being careless*: many organizations attempt to apply agile concepts into projects the wrong way, in the wrong context, and for the wrong reasons—leading to uncertainty, confusion, and *wild-west* behaviors where team members careen forward irresponsibly, attempting to do things too fast, moving forward without planning, making reckless decisions, and causing more harm than good.

OVERVIEW OF AGILE CONCEPTS

Agile is a pragmatic approach that recognizes that there is a lot that the team and the customer do not know, a lot the team and customer learn along the way, and that things change without warning. Agile is a flexible approach that allows us to adapt and learn in real time as things happen, to respond to change, and to satisfy the customer as much as possible within the project constraints and realities. Agile projects start with a clear vision and product and are then broken into releases and iterations. The iterations should be of a consistent length (commonly, but not necessarily, between two and four weeks) and they are grouped together (any one or more iterations) into a release. Every iteration has a theme and goals and is made up of a slice of the project backlog (which is the total scope of the project—consisting of features, stories, and requirements). By the end of the iteration, the team needs to produce a working, production-quality product increment as if it was going to the customer. However, the product increment that is produced at the end of the iteration is for feedback purposes—the final product is actually delivered to the customer at the end of a release, with one or more iterations grouped together (and integrated through regression testing)—so this *chunk* of working product is a fully functional slice of the total product.

Known as iterative and incremental development, agile concepts help realize the following benefits:

- Adapt to changing needs, conditions, and environments in real time
- Produce benefits earlier through periodic releases of product increments
- Improve quality by giving the customer the chance to review progress periodically and improve based on actual performance and on the feedback provided—known as progressive requirement elaboration
- Reduce risk of building the wrong product and building the product with defects (the wrong way)
- Reduce scope efficiently through prioritization and reprioritization—agile enables the efficient reduction of scope by removing lower priority scope items in the later stages of the project once the majority of the value has already been delivered to the client

Additional characteristics of agile include accepting ambiguity in earlier stages until detail is needed, and ensuring that feedback is accepted as soon as possible so the team can properly act based on the feedback.

Lean Concepts, Wastes, and Principles

Additional agile concepts include the following:

- *Continuous improvement of processes (learning)—improving processes and practices on an ongoing basis*: This concept is inspired from the term Kaizen in Japanese (Zen = good; Kai = change). The continuous improvement refers to small, incremental improvements to be learned, realized, and applied on an ongoing basis, unlike concepts such as Kaikaku (from Japanese—radical change) which refer to dramatic and quick improvements. Continuous improvement takes place in small doses over time, which improves the ability to apply them one by one, reduces resistance, and helps isolate the impacts of the individual improvements.
- *Customer focus*: This is about producing and maximizing value for the customer.
- *Lean principles*: The lean movement started in Japan in the mid-1950s in manufacturing (the automotive industry) with its main focus on loss reduction and sustainable production. Figure 1.1 lists the *original* seven wastes identified in the Toyota Production System by Shigeo Shingo. Agile is one of the most significant and successful attempts to translate lean benefits from manufacturing to software development and, subsequently, to other service and non-manufacturing projects. Lean development can be summarized by the following concepts that are very close in concept to lean manufacturing principles:[1]

  1. *Eliminate waste*: Activities that do not directly add value to the finished product are waste. The three biggest sources of waste in software development are the addition of unrequired features, project churn, and crossing organizational boundaries. To reduce waste, the development teams need to self-organize in order to optimize themselves for the work they are trying to complete.
  2. *Build in quality*: There is a need to design quality into the product, instead of relying on later inspection. But when this is not possible, there is a need to break the work into small and more manageable cycles that involve work-validate-fix-iterate. Inspecting after the fact and queuing up defects is reactive and ineffective.
  3. *Create knowledge and amplify learning*: Promote strategies, such as iterative development, that help teams discover what stakeholders really need and act on knowledge as fast as possible. Leaning also involves frequent reflections and improvements.
  4. *Defer commitment (decide as late as possible)*: This is not a call for procrastination, but there is no need to start building a product by defining a complete specification. The design and architecture

need to be flexible so they can change as needed. Deferring commitment to the last, most responsible moment allows the team and stakeholders to make the most informed decisions.

5. *Deliver quickly* (*deliver as fast as possible*): Quickly does not mean reckless; it is possible to deliver high-quality products quickly by identifying team capacity and limiting the work to this level. By becoming aware of its capacity (velocity), the team can establish a reliable and repeatable flow of work.

6. *Respect people and empower the team*: Sustainable advantage is gained from people who are engaged and thinking, and it is achieved by enabling and motivation, rather than controls and limitations.

7. *See and optimize the whole*: To achieve an effective solution, we must see the whole and the bigger picture, including understanding the high-level business processes that individual projects support across multiple systems and areas.

The 7 Wastes

Overproduction
1
Over-processing 7
Inventory 2
Rework 6
Waiting 3
Transportation 5
Motion 4

Figure 1.1 The seven lean wastes of software development: the seven wastes were identified in manufacturing in the Toyota Production System by Shigeo Shingo, then *translated* to software development by Mary and Tom Poppendieck

See Table 1.1 which illustrates the seven wastes of software development.

Table 1.1   Lean wastes: Shigeo Shingo (1981) identified seven wastes of man-ufacturing and Mary and Tom Poppendieck (2007) translated these into waste in software development. For an agile BA with keen eyes toward eliminating waste, this list offers a gold mine of opportunity

The Seven Wastes of Manufacturing	The Seven Wastes of Software Development
Overproduction	Partially done work
Inventory	Extra features
Extra processing	Relearning
Transportation	Handovers
Motion	Task switching
Waiting	Delays
Defects	Defects

Combating Waste

To better understand agile principles and concepts, it is important to review the ways lean attempts to fight waste. For that, we need to check the underlying reasons behind the waste and think about ways to reduce or eliminate that waste. Although this book refers to the benefits of a BA, or the application of business analysis skills in any type of project environment, the wastes we cover here are the translations by the Poppendiecks of the manufacturing wastes that are identified by lean principles to software development wastes. The following are the seven wastes, along with the potential causes and solutions.

1. Partially Done Work

This is work—including stories, features, and requirements—that is not done to the extent that qualifies them to be called *done*. Chapter 7 of this book discusses in detail what *done* means—but in short, *done* means that a feature, functionality, or work product does what it is supposed to do, to the extent that it needs to perform. Defining what constitutes as *done* in agile projects is critical for agile project success. Products and features that are not *done* cannot be released for the client to use.

Potential Causes for Partially Done Work:

- Issues with planning that compromise the team's ability to deal with technical issues, complexities, and dependencies
- Problems around prioritization and the information about the story

- Inconsistencies about following through with the iteration's mandate; removing and adding stories throughout iterations
- Lack of coordination within the team that leads to delays and bottlenecks
- Problems around task identification and matching of tasks to team members

Ways to Reduce or Eliminate This Type of Waste:

- Product owner involvement
- Proper prioritization and reprioritization for each iteration
- Ensure sufficient planning and estimating of stories and requirements (by the team members who actually perform the work on the respective stories); this will also help reduce *task creep*, where the actual work to complete a task ends up being more than the planned effort
- Follow through on the iteration plan and do not change stories (including adding and removing stories) midway through the iteration
- Since planning and prioritization includes managing story dependencies (including dependencies on external elements), ensure the product owner is involved in all stages of the prioritization and when the product owner lacks technical understanding, introduce a BA to support the effort and enhance the coordination and collaboration
- Cross-functional teams can help reduce roadblocks when other team members face problems

2. Extra Features

Extra features are also known as gold-plating, which is a type of scope creep—in short, it is about producing more than what we have been asked to do. Many people and even organizations confuse the term *going above and beyond* with *producing above and beyond*. Going above and beyond is about making a sincere effort to satisfy the client's needs. However, the intent is to satisfy the client within the scope of the work that was agreed upon, and it is not about producing more products or features. Gold-plating is about giving the client more than what they signed up for. As an example, let's think of a chair manufacturer that was requested to deliver a certain number of office chairs. The contract specified the number of chairs, models, trims, finishes, colors (all under the umbrella of scope), costs, and timelines. When the chair maker delivered the chairs, the customer was surprised to see that the chairs that were delivered included the same features of a higher-end and a more expensive model than what was requested. This means that the chairs had more features and were made of a more expensive (and apparently more refined) upholstery. The chairs arrived on time and there was no extra cost to the customer.

Should the customer be happy? The answer is no. The customer asked for the specific chairs for a reason, since the chairs that were requested addressed the customer's needs. Any additional features were not required, asked for, or welcomed. Because the extra features were not in the contract, it may be that they were not designed or installed properly, which will potentially deem the chairs as unusable. It may also mean that the extra features are not appropriate or useful in the customer's environment. For example, it is possible that these are additional chairs to an existing office, and by bringing in new chairs, it may cause conflict in the office as to who will receive the new, improved chairs. Although the customer was not asked to pay more for the chairs, it does not mean that the customer views these chairs as being of a higher value. Further, the delivery of the higher-end chairs may set a precedent in the eyes of the customer that in the future there will be an upgrade every time. Finally, if the customer continues to deliver high-end chairs for lower prices, they may incur losses and may go out of business.

Any way we look at it, there is no benefit in delivering more than what the customer asked for. If the chair maker wanted to do it the right way, they should have asked the customer whether they were interested in the special deal of the higher-end chairs for a lower price. Although this example is simplified, gold-plating happens often and in many cases is not realized until it is too late. Gold-plating can occur on the team level, where team members may decide of their own volition to add features, thinking it might be a good idea to build something in addition to what was required, or to apply their understanding or logic to interpret customer needs. It is clear that it is all in the name of delivering customer satisfaction; but if there is a situation where more features or more things can be produced for the project's work, it is important to communicate and handle it properly by ensuring that there is customer acceptance and understanding of all the related implications.

Team members' awareness of business analysis skills can help in these situations so the team members will perform the work as agreed upon and use the proper channels for any potential additional features. Alternately, a BA is simply another set of eyes to ensure that the work performed is in line with what was required and requested, and that any opportunity for additional features is communicated, addressed, and goes through the proper channels for due diligence, approval, prioritization, and planning.

Potential Causes for Extra Features:
- Lack of clarity around the product vision
- Issues with prioritization of features
- Gold-plating
- Lack of understanding of the true customer needs or product use
- Lack of discipline among team members

Ways to Reduce or Eliminate This Type of Waste:

- Assure product owner involvement throughout—provide clarity around the product vision, reiterating the vision and any adjustments on an iteration by iteration basis
- Communicate the nature of the value to be produced and create awareness around the return on investment of features within the project
- Provide clear prioritization and reprioritization of stories and features; Chapter 7 of this book presents a detailed review of the stories' and features' prioritization process, but in short, prioritization has to be consistent and is based on the following criteria:
  a. Business value (provided by the product owner)
  b. Technical considerations and dependencies (identified by the team and addressed through discussions between the team and the product owner)
  c. Logical groupings (identified mainly by the team; may include considerations related to product viability)

Either way, the product owner has the final say about the prioritization of the backlog items.

- Since gold-plating is often originated within the team and for the most part is unintended, prioritization, open communication, and clarification of the value of features is important—teams can ensure these items are properly addressed with the help of a BA, or the application of business analysis skills

3. Relearning

This is the failure to use knowledge that is available to improve the way the team works or to refrain from stumbling into known (or previously encountered) issues and problems. Relearning also involves team members who attempt to solve problems (reinvent the wheel) on their own, instead of utilizing a solution that is already in place.

Potential Causes for Extra Relearning:

- No proper knowledge sharing within the team
- No proper lessons-learned process at the end of iterations
- Teams that are not colocated
- Issues surrounding communication, including agile reporting
- Problems with (and specifically missing) documentation

Ways to Reduce or Eliminate This Type of Waste:

- Colocation is not always possible, but it is proven that colocated teams perform better than distributed teams. When there is no choice, there are additional factors that need to be introduced to make up for the distance: strong leadership, focus on communication for distributed teams, application of sharing and reporting tools, establishment of expectations, and ground rules for the team and other stakeholders. Did we say leadership? This consists of a product owner that is present, a strong Scrum Master, and most likely the need for a seasoned BA since the sharing of the task to apply business analysis skills among team members may not be successful. This type of sharing is challenging with teams that are colocated, and it becomes significantly more challenging in distributed teams.
- Knowledge sharing throughout and at the end of the iteration is important.
- Properly run daily stand-up meetings will be beneficial.
- Provide real time and continuous updates to information radiators, task boards, and other reporting tools.
- Require participation in team meetings and buy-in to the agile processes and ceremonies.
- Develop documentation. Agile is not about having zero documentation: Chapter 5 covers agile documentation in detail, but we will go back to the idea that documentation is and has to be part of an agile project—multiple times. Documentation is the responsibility of the entire team and it has to take place on an ongoing basis, just-in-time.
- Identify areas (physical and virtual) that are accessible for the team to view, share, and discuss ideas, findings, and performance.
- Perform meaningful retrospectives where team members participate and contribute to the effort. Findings should be established by consensus, applied soon after, and reviewed by the end of the next iteration. The BA can facilitate the retrospective meeting, provide feedback and input to discussion items, monitor the application of improvement ideas, share the results, and report on related issues.

4. Handoffs

A handoff is the handing off of work from one person to another. Lean calls for the minimization of handoffs because they create inefficiencies and open the door for mistakes and misunderstandings. This is not a call to have only one person doing the work, but the intent is to reduce the number of unnecessary handoffs to the minimum necessary.

Potential Causes for Handoffs:

- Some tasks and activities require handoffs
- Lack of availability of, and access to, information
- Distributed team members working on the same items
- Planning problems: issues that arise as a result of poor planning—the need to move tasks from one team member to another, escalations, misunderstandings, and lack of team rhythm

Ways to Reduce or Eliminate This Type of Waste:

- Certain tasks have to be performed by multiple team members; each one brings in his or her own unique experience and role.
- With good planning, the need for a handoff, even when there is a need for cross-functional teams, can be minimized. Proper identification of needed cross-functional skills and resources can also help reduce handoffs.
- Distributed teams typically have challenges with handoffs, especially when there are time zone differences. These handoffs lead to a different type of waste—delays. Focus on communication; coach the team to ensure that all team members own their part of the communication; establish practices for proper sending, receiving, and feedback cycles for messages; and monitor how it takes place for potential improvements.
- Try to perform related tasks and work on specific features in one location.
- Proper updates of task boards, reporting, version controls, and availability of tools, such as wireframes and flowcharts, can also help reduce the number of handoffs and related challenges.

5. Task Switching

Task switching is often mentioned in the context of multitasking. Many people confuse the terms and pride themselves for being able to multitask. While it is important to be able to handle multiple priorities simultaneously, this is not the meaning of multitasking. The large majority of people cannot multitask. Multitasking is about doing two things at the same time with the same part of the brain. While (most) people can walk and chew gum at the same time, when it comes to activities that compete over the same part of the brain, it is a different story. Almost any project practitioner, or even virtually any person who has an office job, has gone through the attempt to do multiple things at the same time. The most common example is to dial into a conference call, then muting the line and trying to do work at the same time. It is so common that we often do not give it a second thought, but the question that needs to be asked is whether or not it benefits us. When we do this type of multitasking, we do not get the level of focus that is needed to perform the tasks we have in mind, and at the same time, we do not pay sufficient attention to the call. The result often is that the work does not get done; someone on the conference call asks us a question, but we do not hear it; the second time they call our name, we respond, but with the phone still on mute; and on the third attempt, we finally unmute the phone and ask them to repeat the question. By now, our train of thought is cut, and as for the conference call—we do not know what is going on there, and the other participants are most likely irritated at us by now for the delay and distraction we are causing.

This was just one example, but it has been proven both scientifically and practically that multitasking not only does not help productivity, but rather hurts it. A study by the American Psychological Association[2] found that multitasking reduces task efficiency and is especially hard on our brain power when we switch to difficult or unfamiliar tasks. It is estimated that we lose up to 40% of our productivity when we multitask, and every interruption *costs* us an additional period that ranges between five and 25 minutes—in addition to the time it takes us to attend to the interruption. To illustrate, if someone is focused on writing a report and an e-mail pops up on the screen, once this person attends to the e-mail, it will take him/her up to 25 minutes after finishing with the e-mail to get back to the level of focus and rigor he or she had prior to the interruption. When adding it all up, how many interruptions can we handle in a day before our productivity becomes a write-off?

When working on an important task, it makes sense to have a second task to work on in case we hit a roadblock with the *original* task. It can help improve our productivity to have a second task lined up to work on, so we do not sit idle while we wait for the first task to resume. However, from here, the law of diminishing returns kicks in rapidly—once we add additional tasks, our productivity plunges significantly. When we have multiple items on our plate (and we all have multiple things to do), it is important to identify our own capacity so we know our limits and what we can do within those limits. Once our work volume reaches the limit, it is time to prioritize our work and determine the sequence by which we will perform the work. We should then focus on one thing at a time based on the priority that is set and proceed with it before moving on to the next task.

Discussing personal productivity, capacity management, and prioritization may not appear to be directly related to agile projects, but these items are in fact key success factors in agile projects. Teams must determine their capacity (measured by velocity), stories and features must be prioritized, work must be capped at the capacity, and team productivity must be maintained at the expected pace sustainably, over time. The BA, along with the Scrum Master, should be both role models in productivity and coaches in applying these concepts (productivity, capacity management, and prioritization) for team members. At the same time, the BA and the Scrum Master should also facilitate the prioritization process and the process of determining and maintaining velocity.

Task-switching becomes an even bigger problem when team members move from one task to another without completing tasks properly during the process, leaving a lot of work in progress (WIP).

Potential Causes for Task Switching:
- Interruptions are a major cause that leads to task-switching. Interruptions are the result of a variety of causes including poor planning, changing priorities within the iteration, a mismatch between resources and tasks, and performance issues.
- Lack of coordination between the product owner and the team. This could be driven by a product owner who is not sufficiently involved or due to communication and reporting issues
- Shared resources when the team works on more than one project. An important success factor for agile projects is having a dedicated project team. While it is possible to share resources with other projects, it often leads to problems and delays since the project would then be exposed to problems, both from within as well as to any issues that hold the resources in the other project.
- Planning issues, such as when task estimates for the iteration are not done properly and the team struggles to meet its commitments.

Ways to Reduce or Eliminate This Type of Waste:
- Make the team dedicated.
- Ensure proper planning, estimating, and sufficient breakdown of the stories into tasks. Also, it is important to ensure that team members match themselves properly with tasks. Sequencing of the tasks is also important in order to allow stories to finish without bottlenecks and with a minimal number of task-switching incidents.
- Reduce the number of interruptions by ensuring that all information about tasks and priorities is clearly communicated and understood, and that tasks are estimated and detailed sufficiently. In addition, identify in advance subject matter experts (SMEs) to be accessed for each story, as well as external dependencies. The BA and the Scrum Master need to facilitate these areas and work with the team to minimize these areas of impact.

6. Delays

Delays slow down the team's value creation and lead to frustration-causing further distractions and interruptions to the workflow. Delays are events, problems, misunderstandings, and slowdowns that cause more time to deliver or to start the work on value-added activities.

Potential Causes for Delays:

- Process issues—when there are bottlenecks and misunderstandings in processes.
- WIP—when there are too many things on the go, or in other words, too much work in progress. This can be driven by performance issues, by unrealistic expectations, or by poor planning.
- Resource sharing and other external dependencies—the team cannot move forward with its tasks due to lack of resources or an extensive need to coordinate the work with external stakeholders and team members.
- Uncertainty—meaning when there is a high number of items that require clarifications, when many impediments and hurdles introduce themselves, and when there are many mismanaged assumptions or assumptions that are not validated.

Ways to Reduce or Eliminate This Type of Waste:

- Efficiencies—identify only the processes that are necessary for the team to produce value for each iteration. Anything that can help reduce the cycle time of tasks and activities should be considered.
- Identify external dependencies and support needs in advance (for each iteration)—it is important that SMEs, stakeholders, and other support functions from outside the team are readily available in a timely manner. This requires careful planning and the ability to identify needs at almost exact timing. The need for proactiveness and coordination is further reinforced with the distributed team; and it should be supported by accessing the supporting resources in advance, if possible, and by always having a backup plan and contact information in case there are connectivity and communication issues.
- Prioritization—beyond story prioritization, it is important for team members to properly prioritize their tasks. Effective prioritization will help improve the overall planning process and allow other team members to pick up work that is based on their skill levels and capacity, helping their colleague where and when there is a need for such help.
- Resource availability—ensure that all required skill sets are available and assigned to the project when required.

- Check processes, performance, and efficiencies—a value stream mapping exercise can help in determining what is value-added time and what is not in order to help improve cycle time.
- Automation—automate test cases and deployment where possible (while considering the cost-benefit) to help reduce cycle times significantly. This is not a call to automate all test cases, but rather to automate those that are recurring and can be reused. Chapter 8 of this book elaborates on the area of testing and provides important information regarding testing automation considerations.
- Manage assumptions—when asked, most people say that they do not like assumptions. That is despite the fact that we all make assumptions. Furthermore, we heavily rely on assumptions when we plan. Managing assumptions means that when we make one, we track it in an effort to validate it. When an assumption is not validated (whether there is no answer or the assumption does not end up playing out the way we assumed), it becomes a risk and should be managed systematically, the same way we manage other risks.

Table 1.2 illustrates how to manage assumptions by providing three columns: (1) what we need to know; (2) who needs to provide us with information; and (3) when we need to know it. Managing assumptions (typically the Scrum Master with support from the BA) can make a significant contribution to the team's ability to handle situations, and proactively identify risks and address issues.

Table 1.2 Project assumptions: three columns that if identified, tracked, and managed properly, can dramatically improve the project team's handle on issues and risks, and help the team become more proactive

Assumption: WHAT we need to know	WHO needs to provide the information	BY WHEN do we need to know it; if the answer is unfavorable, or there is no answer by this deadline, it becomes a risk

7. Defects

Defects (also known as bugs) are errors in a product's functionality that lead to the production of the wrong result. Defects are the epitome of waste since they are produced despite all the efforts to design, build, and test a product. The waste consists of the need to undo and redo the work, increasing costs, and inevitable delays which compromise the value that the team produces, thereby reducing customer satisfaction. All these considerations come in addition to the direct impact of the defect, which may cause multiple problems and issues.

Potential Causes for Defects:

- The story is not a good story—stories need to be independent, negotiable, valuable, estimable, small, and testable (INVEST) in order to be considered as *good stories*
- Stories have no acceptance criteria
- Lack of sufficient technical standards
- Issues around understanding stories
- Insufficient available skills among team members
- Insufficient or late involvement by testers
- Not enough testing automation

Ways to Reduce This Type of Waste:

- Establish a clear definition of *done*
- Follow the INVEST principle
- Break stories down sufficiently
- Clarify the acceptance criteria for each story with the product owner
- Prioritization to ensure that the team's effort works toward producing value
- Ensure the right skills are available on the team (and in a timely manner)
- Establish relevant and sufficient coding standards and guidelines
- Ensure clear understanding of stories prior to the start of work on these stories
- Follow agile principles—in this case, simplicity, in order to maximize the work undone
- Follow agile practices that are suitable for the project's needs, the organization, and the team
- Involve the testers from the start—ensure the testers are part of defining acceptance criteria for each story and that the test cases are written based on the acceptance criteria
- Testing automation

Waste Summary

The wastes, causes, and proposed high-level solutions discussed here are part of our introduction to agile concepts. Chapter 2 of this book takes a deep dive into agile challenges and into ideas on how to address these challenges. The items discussed in Chapter 2 are directly related to agile projects, while the items discussed in this section are addressing only the lean wastes. Although aligned, they are not all the same.

Back to Agile

In February 2001, members from XP, Scrum, DSDM, and other methods got together in a ski resort in Utah—including luminaries Kent Beck, Ken Schwaber, Alistair Cockburn, Jim Highsmith, and Steve Mellor—and decided to each keep their respective methodologies and their characteristics, but put them all under the same umbrella and call it *Agile*. They also produced a set of value statements about what their methodologies stood for and they named it the *Agile Manifesto*, shown in Figure 1.2. It is important to read the Manifesto as it is intended to be and not out of context; keeping in mind the very important statement at the bottom: *while there is value in the items on the right, we value the items on the left more.* The items on the right are foundational, and critical for success, but the items on the left are the ones that will take us to the next level. There are many ways the BA can help in delivering on the value statements expressed in the Manifesto, so let's take a quick look at the benefits the BA can add toward achieving those values by helping to attain the items on the left, but also ensuring that the items on the right are in place:

- *Individuals and interactions over processes and tools*: Processes and tools need to be in place to provide some predictability and familiarity with how to conduct the work and what to expect. The BA can help provide the team with access to processes, information, and context; help improve and streamline the process; as well as facilitate adjustments in application of the processes in the agile project environment.
- *Working software over comprehensive documentation*: While the original intent of agile was to improve software development, agile has grown to be applied in essentially any type of project, so we will keep referring to concepts in the context of any type of product, and not only software. With that said, it is not easy to produce a working product and there still has to be sufficient documentation and development of prototypes, wireframes, or proof of concepts. The BA can help facilitate these elements and provide support for the team to maximize the value they produce.
- *Customer collaboration over contract negotiation*: It is hard to achieve true and meaningful customer collaboration, but it is important to note that the need for customer collaboration is important in any type of project life cycle—agile or not. Sticking to the word of the contract rarely yields the desired results without working closely and collaboratively with the customer in order to truly understand their needs. BAs were originally introduced to improve customer collaboration; their role is to ensure that it takes place and to facilitate the process by providing the team with the necessary support and enablement.

• *Responding to change over following a plan*: This is one of the most important aspects of agile—the ability to respond to change effectively, in real time, and at any time throughout the project. Business analysis skills are important to ensure change accountability, so when the scope changes, timelines need to change too. With the concepts of time-boxing, an iteration (and potentially, ultimately, the project) should not get extended, but the team needs to have the ability to account for changes and to identify the impact of these changes. This takes us back to the need for capacity management (measured by velocity) and prioritization. Ideally, team members should be able to account for the impact of changes, but depending on the project context and the specific business analysis skills of the team members, a BA is often required to help ensure that all scope changes are accounted for.

The Agile Manifesto

Individuals and Interactions	over	Processes and Tools
Working Product	over	Comprehensive Documentation
Customer Collaboration	over	Contract Negotiation
Responding to Change	over	Following a Plan

That is, while there is value in the items on the right, we value the items on the left more.

www.agilemanifesto.org

Figure 1.2   The Agile Manifesto: keep in mind the very important statement at the bottom

Agile Principles

Agile project characteristics include early and continuous delivery of usable deliverables through short delivery cycles and simplicity. Agile recognizes that usable deliverables are the best measure of progress and of value creation for the customer—and it accepts that requirements change, even late in the project. In agile projects, business people are involved daily with the project team to maximize the use of face-to-face conversations, which sometimes introduces challenges. While there is no reliance (like in a waterfall project) on the BA to ensure such communication between the business and the team takes place, a BA may need to facilitate certain elements of these interactions.

Team members in agile projects are motivated, trusted, experienced, and self-organizing. While that is a good thing, many people tend to misinterpret it by envisioning projects with no BAs. There is obviously nothing wrong with having teams that are *qualified* according to all of the descriptions mentioned, but in reality, many teams fall short of these qualifications, and this is where BAs can and need to become part of the project. In fact, for the team to be self-organizing, most team members need to possess and demonstrate business analysis skills. Ideally, agile projects should not need specific individuals whose titles are BAs, but the circumstances surrounding many projects dictate that need. Our goal is not to force the use of BAs in projects, but rather to recognize when a BA is needed and to what extent. The opposite also applies: while agile projects are more likely to get by and succeed without a dedicated BA (than waterfall projects would), we should not force the removal of BAs from agile projects just for the sake of having no BAs in and of itself. BAs can also facilitate the team's learning process through the application of lessons learned and the ongoing focus on continuous improvement.

Tools and Techniques

Let's review a few high-level tools and techniques that are relevant for agile projects. These will be discussed and elaborated on in detail throughout the book:

- *Agile planning*: Agile values planning and the act of planning more than the plan that is being produced. It is important to differentiate between the two aspects. While waterfall is big on the plan that is produced at the beginning of the project, agile is about value-creation and planning on a high level up front—with details being figured out just in time. Planning, monitoring, and adapting to change are about creating manageable and lower risk cycles (following the Deming Cycle of Plan-Do-Check-Act and focusing on the concept of rolling wave planning where we plan a little, do a little, learn, and adjust).
- *Agile estimation*: Like planning, estimating is also about the activity itself, which is a team-building activity. There are estimating techniques that are engaging and they are performed in such a way that it increases the chance of producing realistic estimates. There are also multiple checks

and balances around estimating (and planning) in order to move from high levels to lower levels and to refine our understanding of the needs. Chapter 6 of this book focuses on agile estimating.

- *Ambiguity until further information is required*: We need to plan only on a high level and provide details only for the short-term planning horizon. Simplicity and value-creation are key, but it is not about procrastination—deciding later means that by then, we will have a better understanding of the solution (the big picture) and how the requirement fits in, there will be fewer unknowns as uncertainty is resolved over time, and we will end up spending less time on understanding requirements that will change or be canceled anyway. However, we must keep in mind that we still need at least a high-level set of requirements to help prepare an overall solution outline/model.

- *Communication*: Agile relies heavily on clear and effective communication that is delivered directly, timely, and with transparency. It includes reporting mechanisms (in real time, called information radiators) to efficiently and clearly convey to all stakeholders the most recent and accurate set of information.

- *Interpersonal skills*: With the transparency and the ongoing involvement of the customer, there is a need for strong interpersonal skills and the ability to manage ambiguity, stress, uncertainty, and conflicting priorities. While the role of the Scrum Master (or PM) is to protect the team from external interruptions and to lead the communication process with stakeholders, the BA can help create a bridge between the customer and the project team, as well as within the team and with internal functions and SMEs.

- *Continuous process improvement*: The team should learn from lessons in real time—every iteration. We do not have the luxury to wait until the end of the project in order to apply improvement initiatives.

- *Product quality*: Every iteration produces a production-quality product increment that is ready for shipment to the client. It does not mean that the client will want to get a product increment at the end of each iteration, but whatever we do must minimize WIP and be as good as if it is going to be delivered to the client. Iterations are grouped into releases of meaningful product increments that represent a critical mass of value for the client that justifies the effort to deploy and use the product in the interim.

- *Time-boxing*: Iteration (or Scrum sprints) are time-boxed, which means that they have a preset duration. This duration should reflect the project and customer needs, the nature of the product, and the team's velocity (its rate of producing value). Once we understand our velocity, we can determine how many features will fit into an iteration, and by multiplying the number of iterations, we can provide a fairly accurate time frame for finishing the project.

- *Risk management*: Agile focuses on reducing risk—the risk of building the wrong product and the risk of building the product the wrong way (with defects). Also, thanks to checks and balances, there is a better ability to identify early on how we progress; and if the progress is not satisfactory, it is better to realize it early on rather than later.
- *Value-based prioritization*: Prioritization of stories (as well as requirements and features) is critical for project success. It is done and determined by the product owner and the most important factor in the prioritization process is whether an item represents business value.
- *Burndown charts*: Progress is measured through burndown (or burnup) charts. These are graphs that show the plan versus actual velocity, pace of value creation, and completion of stories. It is a clear indication of the rate by which we progress toward finishing the project and it provides an opportunity to act proactively and in real time on changes in the velocity or the needs.

THE BA, EFFICIENCIES, AND AN EYE FOR WASTE

Business analysis skills are a key to project success; no matter what type of project or life cycle we manage it by. The premise of agile principles is that team members need to have and demonstrate business analysis skills, including the following elements:

- *The ability to pick up stories*: As part of being self-directed, agile calls for team members to assign themselves (or volunteer) to work on stories (requirements) that are in the backlog. It should not be part of the role of the Scrum Master, or coach, to assign work to team members since there will be more context and buy-in if team members pick up stories to work on based on a combination of the project's needs, priorities, story complexity, experience, relevance, the applicable skills and experience of team members, and other things that need to be done on the project. Having team members consider and act on what is the best story to work on requires them to have a lot of context and understanding of the big picture. Having people picking up work on their own provides an additional benefit—enhanced buy-in—as it is likely that team members pick up items that they desire and are qualified to perform.

While it is not implied that technical team members lack such skills, it is important to keep in mind that in many environments, team members will not have the time or capacity to perform such due-diligence analysis—that essentially optimizes the story selection to all other moving parts around us. For situations where team members cannot pick stories on their own, there is a need for a Scrum Master/PM that will help facilitate this process; and this is typically done through collaboration with the BA, who often has additional insights, context, resource availability information, prioritization consideration, and other relevant decision support information.

When the "fathers" of Agile got together in 2001 and named the three agile project roles (i.e., product owner, Scrum Master/coach, team member), they were adamant about not including the PM as one of these roles. It was important for them to differentiate agile projects from *traditional* ones by excluding the PM from it so that they would not imply that the role of the PM is about bossing people around, putting out fires, policing team members, or *herding cats*. With a set of technically excellent, self-directed team members, it was an effective way to differentiate the characteristics of an agile project from that of waterfall projects: the role of the Scrum Master/coach is to add value by protecting the team and being a servant-leader, rather than being there as a safety net for a team that does not perform to the extent that it needs to.

While the intent behind the *absence* of the PM role from the three agile roles is clear, the same cannot be said about the role of the BA. There is a strong need for business analysis skills, and if these skills are not available across the team, there is a need for a BA to complement the situation.

- *Coordination and collaboration*: We need to consider these items in more than one context—coordination with colleagues, within the team, across the organization, with SMEs, with stakeholders, externally (with vendors), and with the customer/product owner. It involves clear prioritization, the escalation process, clarifications, estimating, the impact on other areas, story selection, task identification, changes to performance, and changes in needs. There is a need to ensure seamless collaboration between the developers and the testers, between the developers/testers and designers/architects, and between those who identify requirements and the rest of the team. With fast-tracking being common (doing requirements and design work one iteration in advance for the next iteration), it is often not sufficient to leave all the coordination and collaboration in the hands of the team members, and it may require a BA to facilitate some of these efforts. There are a lot of moving parts in agile projects, and even though there are not supposed to be changes to the scope midway through an iteration, team members may end up being too focused on their tasks, rather than on the handoffs, coordination, and collaboration.

- *Communication and reporting*: Agile projects introduce a different way to communicate and report on project performance. There is a lot more transparency, real-time information that flows, and direct contact between stakeholders and the team. Naturally, these conditions may take some team members out of their comfort zone since they may not be comfortable with the real-time visibility into what they work on at any given time, with the customer's involvement, with the constant need to provide updates, or with the less formal ways of reporting and communicating. For many, formality and documentation provide a sense of security and comfort, and the agile project typically takes that comfort away from many team members and stakeholders. When people do not feel comfortable with the processes around them, they may either take shortcuts or default to old ways of doing things. The impact on other people may be in the form of stress and performance issues. Most of these issues present themselves when team members lack business analysis skills or when people run out of time or capacity to demonstrate their business analysis skills.

  In uncertain situations, people tend to focus on their primary role and push aside secondary chores they need to perform; and with business analysis skills falling under the latter category, those will be the first things to go. Here, too, the intent behind having a BA within the team is not to pick up slack that people neglect to perform or that is a result of carelessness, but rather to complement team members' skills and provide them the necessary support at their weak points so they can focus on their core role and maximize the value they produce.

- *Eye for waste and process improvement*: One of the core traditional roles of the BA is to keep an eye on waste and to identify and implement ways to improve processes. Earlier in this chapter we discussed the common types of wastes identified by lean and we saw how the BA can help alleviate these situations, especially when things change in the organization and when teams transition into agile projects. Team members may not have the capacity or priority to focus on efficiencies, waste reduction, and continuous improvement. While these things are important, they are not always tangible, and they are not part of the core role of any team member. As such, similar to the other items discussed in this section, team members will scramble to produce the work products associated with their core role. They will then focus on delivering tangible deliverables (or work products) and only then (typically the last item on the list) will they get to the efficiencies and improvements. With these items being part of the core roles of a BA, we should expect the BA to virtually always pay attention to these items. Another argument for the need of a BA is that leaving the improvements activities in the hands of many may end up where no one pays true attention to them, and as a result, improvements may never take place—at least not until they become necessities, pains, or problems.

Organizations must check whether there is sufficient capacity for team members to perform business analysis activities—or identify someone to do that for the team. Just by labeling a project as agile, we do not guarantee that everyone will do exactly what is expected of them, especially when it does not go beyond their core traditional role. Part of the continuous improvement activities of the team is achieved through the retrospective where it is important to capture, document, and apply lessons learned, as well as to track and report on the extent of the improvements. When the team does not have the capacity to do so, a BA needs to facilitate these activities.

- *Adapt to change*: One of agile's core intents is to handle change and this is done through communication and collaboration with the product owner and the activities around backlog maintenance. A few situations may introduce the need for a BA: (1) when the product owner is not sufficiently available, a BA can act, partially, on behalf of the product owner (though not for determining business value); and (2) someone (or more than one person) within the team needs to capture the feedback from stakeholders that is received during the demo and translate it into actions and updates of the project backlog (this is backlog maintenance). Further, once there is approval for the changes on the backlog, team members need to move forward with the planning, task identification, estimating, and coordination process. When team members do not have the capacity to go through these activities, once again—a BA can help.

RECAP: AGILE CONCEPTS

Agile methodologies are designed to overcome many of the challenges that *traditional* project life cycles enable (e.g., waterfall or predictive life cycles). Although there is no clear mention of the need of a BA in agile projects, in many environments having a BA as part of the team may be one of the keys for success.

First, by trying to overcome problems and do things differently, agile projects change the way teams do their work, introduce efficiencies, and require team members to demonstrate business analysis skills. Where team members lack these skills or where processes do not enable the demonstration of these skills by team members, there will be a need for a specific BA role.

In addition, like any other type of improvement, agile projects introduce new challenges that if not addressed may become issues and problems. These new challenges are triggered as a result of doing things differently and they may appear as *side effects* of the team's performance. They can range through pretty much anything—from estimating challenges, coordination, the ability to see the whole picture, ambiguity, different (and typically reduced) documentation, or changes in reporting and communication. Whatever it is, a BA can help in closing the gaps and cracks these issues may create while complementing the skills and experience of the team members in the process.

Whether the intent is to utilize the business analysis skills among team members or to introduce a BA role to be part of the project team, the intent of agile business analysis is for a more enhanced business analysis role that complements relevant skills of the team—as opposed to some of the older views of the BA as a bridge and a translator to make up for skills deficiencies, communication breakdowns, or lack of attention to detail by team members. While ideally the agile team should be comprised of technically excellent, motivated, empowered, and self-directed team members, in reality, the setup is not always sufficiently conducive to achieve the project's goals and at times it needs to be enhanced by ensuring business analysis skills are demonstrated—one way or another.

There is no one magic way of looking at things to determine whether there is a need for a BA in a project or that the team has what it takes to cover the business analysis activities and areas. Similarly, some of these skills may be filled, fulfilled, or supported by the Scrum Master, coach, or PM. However, we must make sure that the person leading this project has the knowledge, capacity, and context to do so—without compromising other aspects of their role.

ENDNOTES

1. http://www.disciplinedagiledelivery.com/lean-principles/.
2. http://www.apa.org/monitor/oct01/multitask.aspx.

2

AGILE CHALLENGES

Agile is not a fad; it's not going anywhere. However, there are many misconceptions around agile; not only about how to adapt to agile methodologies and what practices and ceremonies we need to perform, but also about what agile is really about. There is confusion around what agile is (an approach or an umbrella) and with it—confusion about the methodologies under this umbrella (i.e., Scrum). There is also confusion about what it takes to be agile. With the evolution of agile methodologies, new terms and names are added to the mix [for instance, Disciplined Agile Delivery or Development and Operations (DevOps)], and with them more misconceptions, mixing and matching, and above all—cherry-picking. People watch 10-minute videos on the internet and think they are now experts in agile, while others just pick and choose concepts they hear about and try to make them work.

The problem, though, is not the misconceptions, but rather what the misconceptions lead to—confusion. This confusion translates to all sorts of attempts to bring in agile methodologies and there is often a mismatch, an attempt to apply the wrong concepts and practices, and confusion that leads to poor results. The next step is to blame agile for their failures even when this is not the real reason.

Many people use the terms *capital A Agile* versus *small a agile*; trying to differentiate between following agile methodologies, as opposed to becoming more agile. Many organizations embark on attempts to bring in agile methodologies for the wrong reasons; stakeholders and team members often do not know what agile is really about and their goal is simply to *become more agile*. They want to do things faster, reduce documentation, finish stuff early, and spend less time planning. While agile methodologies offer progress in all of these areas, *going agile* for these reasons will most likely yield unintended results.

Let's take a quick look at these common items listed as part of *going agile* and check where they fit into the big picture:

- *Do things faster*: Agile is not about working faster; it is actually common knowledge that when we work faster, we are more prone to making mistakes. Many organizations are plagued by slow progress in their projects, and senior stakeholders often think that agile is about working faster. In reality, we do end up working faster in agile projects. This is not a hasty or a rushed type of fast, but it is about working for short intervals that produce results in the short term. Working faster is not a goal in and of itself, and it is not an agile goal.
- *Reduce documentation*: Our goal in agile projects is to be efficient, reduce waste, pursue minimalisms (just enough, just in time), and maintain simplicity. This includes documentation. Our goal is not to eliminate documentation, but it is one of the means to achieve efficiencies and to maintain focus on what really matters—adding value to the customer.
- *Finish stuff early*: This item is often related to the idea of working faster. Agile is not about finishing our work early—it is about finishing on time. We do not work faster, but rather we work responsibly. Agile methodologies focus on reducing risk and adding value for the client. As such, we try to deliver value early (through product increments at the end of each sprint and release). Our goal is not to finish the project early, but at times we may be able to efficiently cut the scope by finishing the project before its final deadlines once we deliver the bulk of the intended value. Delivering value on an ongoing basis throughout the project is one of the main benefits agile projects offer. Typically, due to the nature of agile projects and the delivery of value throughout the projects, there is a significantly higher chance that we will be able to complete the project by the deadline. Unlike projects in more deterministic environments that often finish late due to the nature of late feedback cycles and the need for bug fixing, agile projects focus on gradual delivery of value throughout the projects and early and continuous feedback cycles that allow the project to finish on time.
- *Spend less time planning*: It is safe to say that many project environments focus too much on producing plans and project documentation, instead of effective planning.

Beyond the challenges that are triggered by common misconceptions, there are multiple other conditions that pose and introduce challenges in agile projects. These are due in part to organizational challenges and constraints, and in part to individuals' behaviors, practices, and conduct.

DIFFERENT VIEWS OF WHAT AGILE IS

We often hear questions from people to the tune of, "So when does Scrum start?" which indicates a fundamental lack of understanding of what agile is about. This, in turn, leads to challenges since these people tend to look for a series of actions—all tangible—that in their view are part of being agile. While there are all sorts of activities and ceremonies associated with the methodologies under agile, it is important to note that agile is an umbrella, a mindset, and a concept under which there are several methodologies that reside—each with its own practices, ceremonies, and activities.

In a recent project we led for a government agency, we noticed wide differences among stakeholders and team members as to what agile was really about. Some thought it was some sort of a slogan (let's work faster), while others believed it was about changing everything we do. One of the main findings of the project was that the government agency was not ready with its existing practices to move toward agile projects. This did not mean to indicate that they were not capable of managing agile projects, but rather that their practices, processes, reporting structures, and deliverables were all aligned toward waterfall methodologies. When we provided the findings, some stakeholders took it personally and disagreed. They thought we implied that they were not doing things properly and they viewed our work as an audit, instead of viewing it as an assessment of how aligned they were with agile methodologies.

Further, in the assessment process, we noticed a wide variety of opinions as to what agile was about and what they needed to do moving forward. The opinions ranged from *we need to work faster* and *we need to deliver better results*, to *we are so far from being agile, that we are better off starting with DevOps*. These kinds of statements were an indication of how unprepared that organization was to move toward managing agile projects. It was ironic that some stakeholders called for DevOps since the main challenges that organization had were communicating and collaborating among its product management, software development, and the operations staff—the very foundations of DevOps.

The DevOps environment emphasizes cross-functionality, shared responsibilities, and trust by extending the continuous development goals of agile methodologies into continuous integration and continuous delivery. As some view DevOps as a more *extreme* version of agile, it would make sense to start moving toward agile approaches in a more gradual way, rather than diving into the deep end with DevOps.

This is just one example of the misconceptions about what agile is and how to bring it to life. These misconceptions lead, in turn, to picking the wrong approach, applying agile in the wrong context, and conducting the wrong practices.

WHY MANY ATTEMPTS TO GO AGILE FAIL

Projects fail for a variety of reasons, many of which can be fixed by agile approaches. However, when agile is not the suitable way to manage projects, it may cause the projects to fail. Agile offers a series of benefits that reduce risk overall, and if done properly, will increase the chance of delivering project success.

The following list captures key agile principles. These principles serve both as a means to deliver on the *agile promise* for better and earlier creation of value as well as reasons for taking on agile projects:

- *Learning and adaptation*: We recognize that we do not know everything—especially not up front at the very start of the projects. With this recognition, along with the reality that many projects will change throughout their life cycles, agile approaches can help by breaking the projects into small *segments* (also known as releases, sprints, or iterations) that are managed through the concept of *rolling wave planning*. We start with high-level planning (coming up with requirements and estimates) for the entire project with just the details for the upcoming time frame and for the work that is slated to take place in it.
- *Collaboration*: This is a major element in agile environments. Agile projects rely heavily on collaboration at all levels—collaboration with the customer and the user, collaboration within the team, and collaboration across the organization. While easier said than done, in order to achieve meaningful and productive collaboration, a series of steps needs to take place; many of which require the expertise of a business analyst (BA) or at least a business analysis skill set. Collaboration goes beyond sitting in long meetings and creating excessive documentation; it is about working together effectively and recognizing that all parties and stakeholders involved are part of the effort to deliver value for the customer.
- *Customer focus*: Agile projects are all about delivering value for the customer and doing so early on and throughout the project. Value is produced through the creation of features and functionality that are meaningful for the customer. Here, too, creating value involves the need for business analysis skills, ongoing customer and user engagement, an

effective requirements-elicitation process and a meaningful dialogue that truly unveils and focuses on the customer's needs.

- *Self-directed teams*: Although this item is an agile principle, it easily falls under the category of agile challenges. There are many views about the extent to which teams need to be self directed, but one way or another, this is a major building block for agile environments. Being self directed is not only about team members stepping up and helping each other, but it is also about the rhythm and the lifeblood of agile—team members need to pick up stories and tasks to work on of their own volition, without a project manager (PM), a cat herder, or a micromanager assigning duties to them. In fact, when team members are not self directed—including failing to pick up stories to work on, failing to report progress or issues, and working in silos—it is not only a bad sign for the team's agility, but it is also an acute warning sign for the project's overall direction.
- *Technical excellence*: We can safely agree that for the most part and in most project environments, team members have sufficient technical knowledge to perform the project work. With that said, however, it is clear that technical excellence does not just happen and appear out of nowhere. Jim Highsmith once said that *technical excellence in agile projects is measured by both capacity to deliver customer value today and create an adaptable product for tomorrow*. Teams often pursue a simple design that is flexible and easier to maintain, but it may come at a price of limited ability to evolve and meet emerging and expanding needs. Putting together teams that are technically excellent is not an easy task; and while team members definitely know what work they need to perform, there is often a need for a BA to bridge some of the skills and experiences that team members bring to the table. Some agile practitioners view *technically excellent* to mean resources who have a well-rounded skill set that includes business analysis skills and other disciplines beyond their areas of expertise—for example, having software developers who also know how to test. To simplify things, let's view the term *technically excellent* as describing one's strong ability to deliver his or her role successfully and to continuously focus on minimizing technical debt. *Technical debt* is a term that generally refers to features and functionality that were supposed to be delivered throughout an iteration, but for some reason they were not delivered or did not complete properly. Lack of technical excellence is a major contributor to technical debt and it appears in multiple forms: (1) leaving incomplete pieces of work; (2) dealing with overly complex requirements; (3) focusing on dictating

technical solutions instead of introducing solutions to business problems; (4) lacking understanding of how the solution works; (5) leaving
work undone; (6) over-engineering and gold-plating; and (7) cutting
corners about processes and best practices. With a constantly increasing
number of organizations managing agile projects, there is an increasing
need for business analysis skills within the team to bridge gaps in technical excellence. Being technically excellent is also about understanding
that working faster often translates to a slower pace of producing value,
as more defects and issues are introduced and left unattended (technical
debt).

- *Lean principles*: Last but not least, agile environments are about reducing
waste, minimizing work in progress, reducing handoffs, and focusing on
producing value for the customer. These lean principles provide a proper
summary of the intent behind most attempts to go agile, but as previously mentioned, they are not easy to implement or introduce into any
project environment. In summary, these lean principles are also about
being responsible for our actions and accountable for the results that we
produce; gone are the days where technical team members can get away
saying, "The business made me do it."

AGILE, CULTURE, AND ORGANIZATIONAL CHANGE

A very knowledgeable stakeholder in a recent project articulated the magnitude of change that agile introduces to organizations by saying that agile is *a
full-blown organizational change, including everything that comes with it: resistance, processes, and policies. It is not just about the difference between Coke
and Pepsi.* The impact of this organizational change is not merely another
run-of-the-mill challenge that may put a dent in the organization's ability to
realize the benefits agile promises. In many cases it may translate to a full-
blown failure of the project and with it the misconception that agile is the
reason for the failure.

An agile project requires virtually everyone involved with it to change the
way they do things. While the change may not be significant, it is a change that
goes across many aspects of the daily familiar routine of people in the organization. As such, it will trigger resistance and lead to various levels of conformity,
adaptation, and results. The change spans multiple areas, including changes to
job titles, reporting mechanisms, planning cycles, turnaround times, roles and
responsibilities, and above all—change in the way we do things. In many work

environments, there is a culture of *us versus them* on multiple levels: development team members view testers as *outsiders*, there is the traditional *animosity* between the business and the technology groups, and most commonly, there is the culture of closed doors where the development team is performing its work while being *hidden* from the customer and any interaction takes place through a *translator*, a BA, or someone else who takes over the awkward task of being the bridge between these parties.

Agile methodologies focus on collaboration and individuals, but it is not always easy to bring these concepts to life; especially in organizations where the aforementioned behaviors are part of a long *tradition* of doing work.

A Review of the Agile Manifesto

Let's go over the Agile Manifesto and review the magnitude of the organizational change within it and the benefits an agile BA can bring to the table in trying to weather this storm.

Individuals and Interactions over Processes and Tools

For many individuals, their source of power is red tape, processes, and policies. It is hard to shift their paradigms and it is even harder to shift an organization's DNA away from this kind of thinking—but it is critical for the success of the agile process that we shift the way we think and that we let go of these *traditional* but less effective sources of power and focus on individuals and interactions. Focusing on individuals may lead to more risks, misunderstandings, and communication breakdowns, which may lead to less control over the negative impact that certain behaviors have on the team and organizational performance. However, if done correctly and supported by business analysis skills, it can lead to a more nimble, effective, and efficient creation of value. The improvement will empower individuals to grow, while still offering sufficient guidance and controls. This growth will in turn help enhance individual performances and their team orientation.

Agile planning cycles call for allowing ambiguity (in requirements and estimates) until more information is needed, so it will require all stakeholders involved to come to terms with the fact that ambiguity is welcomed and with the fact that project management is not an exact science. Unwelcomed ambiguity may trigger resistance, stress, and at times, panic—but if it is managed properly and done in a controlled way, it will help establish the rhythm of agile and support its effective planning cycles.

Working Software over Comprehensive Documentation

This concept is easy to agree with. After all, who does not want to view themselves as ensuring that their product works, rather than being viewed as heavily focused on documentation? Documentation is part of our lives. It is part of who we are, how we do things, and how we manage to achieve our goals. We can identify a point in time, most likely in the early 1990s, where we started to see the emergence of approaches and maturity models; which with time have turned us into what some might call *documentation monsters*.

These maturity models (e.g., Capability Maturity Model, and subsequently, Project Management Maturity Model) had an important role in building cultures of consistency, continuity, and high performance, and definitely helped ensure that organizations had a backbone of processes, documentation schemes, models, and measurable performance levels. The growth of these models took place along with more rigorous views of process adherence, the wide adaptation of Six Sigma, and the increasing demand for documentation, approvals, and sign-offs. These trends and mechanisms helped organizations achieve and maintain higher service levels and directly helped improve performance. The downside was—and still is—excessive documentation.

We often find ourselves in situations that are brought to a virtual standstill, solely due to excessive documentation and sign-offs. As an example, a few years ago we led a project for a government agency in Canada where it took over eight signatures in order to change a light bulb. The multiple rounds were due to political considerations, inefficiencies, duplication of effort, and *territorial thinking* (referring to processes and steps that are in place due to lack of trust or lack of understanding of the process and by the respective individuals protecting their territories and their areas of responsibility with red tape). The eight-level approval process led to a situation where every time there was an error in the extensive approval process to change a light bulb, the person who discovered the problem sent the application to its originator, leading to further delays and waste. In total, the cost of changing a light bulb in that organization exceeded $1,500 when counting all contributors, approvers, forms, signatures, and rework. As we live in an era that resembles a pendulum, we are now starting to realize that it has swung too far toward being heavy-handed in our documentation—and agile methodologies are trying to address that. There is a place for documentation and for processes, but it has to be recognized that these are a means to achieve results and therefore are secondary to performance, features, value creation, and a working product.

Agile methodologies recognize that we cannot achieve our goals without documentation and processes, but there is also the recognition of the damage that excessive documentation may lead to. We therefore try to strike a balance between the two opposing forces; doing so by trying to reduce documentation to the minimum necessary. It is a common misconception that agile methods go without documentation, so there is a need to combat this misconception as well as to address people's natural tendency to protect themselves with excessive documentation and at times *hide* behind documentation to delay action, shirk responsibility, or hang on to sources of power.

There could also be more *legitimate* reasons for excessive documentation, including the need to reduce risk. Unfortunately, combined with the other reasons, excessive documentation not only fails to reduce risk, but it often increases the risk for not doing the right thing, not doing it correctly, or failing to do it in a timely fashion.

Balancing the need for power, efficiency, and reducing risk can be done with the incorporation of business analysis skills to ensure timely and sufficient documentation, along with the incorporation of prototypes, wireframes, or pilots—if and when relevant. The extra set of eyes, including the natural tendency of the BA to have an eye for efficiencies and for reducing waste, can help ensure we do not take too many shortcuts with documentation and that there is a check and balance mechanism for this process.

Customer Collaboration over Contract Negotiation

This value statement is easy to agree with. Actually, we should take this approach in all projects—agile or waterfall. But there is more to it; the reason we need to resort to the contract in many projects is the nature of the delivery of value in waterfall projects. While there is a place and time for these types of projects, the first chance we have to see what we have in a waterfall project is at the very end. If things go well, the customer will be happy and we will all move on with it. But when there are problems, it is too late to start tracing back to discover what works and what does not work, or why and where the problem originated. Upon reaching the deadline, the customer wants a working product, and if the product does not do what it needs to, to the right extent, the customer will not be in a state of mind to hear excuses or timelines of how long it is going to take us to fix the problem. Instead, the customer will quickly refer back to the contract, which will spell the end of the relationship as we know it.

Agile approaches deliver value early and on a continuous basis. The customer has, potentially, multiple opportunities to realize the value and to see the product in action. Even if toward the end we still have some loose ends and some low priority items that are incomplete, the customer will be much more inclined to work with us toward achieving the remaining pieces of value. Trust, relationships, and rapport are built gradually throughout the project by showing progress in features, functionality, and realization of value, and the customer does not need to take your word for it when you show progress on the status update.

The word *collaboration* has to be taken in the right context and even more important, both the delivering organization and the customer must be on the same page as to what this word means to each other. It is nice to work together and *collaborative* but the customer may have a completely different view of what that means, and it may translate to having a customer on-site for the project team. We should therefore work with the customer and establish a clear understanding of how to realize value, how often the customer wants to be involved, how often releases need to take place, and what the iteration length should be.

Collaboration is also something to be sorted out within the team. A complaint we often hear from some agile team members is that they want some peace and quiet. Those asking for it are often introverts—members of the team who struggle in dynamic, fast-paced, and often loud agile environments. All they ask is for a quiet space in which to do their work.

Whether it is collaboration with the client, within the team, or with other project teams within the organization, business analysis skills can be very handy for these tasks. With the increasing number of moving parts, more frequent interactions, and the changes in individuals' roles, those with business analysis skills can step up and offer support in areas related to the reporting, roles and responsibilities, interactions, feedback loops, and realization of value. This is not a call for the return of the BA as a generalist, or a *master non-generalist*, but rather someone who is the backbone of the team, ensuring things are all synchronized.

Responding to Change over Following a Plan

So, if done properly and effectively, agile addresses one of the most painful aspects of project management: change (in the context of project change control and scope change, not organizational change). Every project practitioner knows that change (along with risk) is one of the most damaging things that takes place in projects. Further, as change requests keep coming in, PMs start having their role distorted and they become *scope protectors*, trying to block changes from taking place instead of allowing them to be accounted for. The reason for trying to block changes is to reduce the impact that these changes have on the project and on the team. The changes introduce instability, uncertainty (in addition to other risks throughout the project), and confusion. The changes consume resources (who now need to assess the impact of the proposed change instead of working according to the project plan) and divert the team's attention from doing the work to planning on the run.

By nature, change requests also increase the scope, and in turn, impact the schedule, resource allocation, costs, and potentially, quality. In addition, the change, once accepted, also impacts the project status. For example, if the project's status shows that it is 50% complete this week, and a change request that increases the scope by 10% is approved after the weekly status meeting, the change basically means that the new status of the project is now only 40% complete. It would be hard—awkward, even—to stand in front of our stakeholders at the next status meeting trying to explain to them that we actually moved backward in the past week.

The impact assessment of change requests is also sketchy at times. We need to perform the impact assessment across the project, while other things are moving forward, and come up with a realistic impact of the proposed scope increase (let's admit that the majority of the change requests in projects should be called *more requests*, since they are not so much about change, but rather about scope increase). The impact we try to measure is what it will take to perform the change looking at all the other aspects of the project: schedule, resources, costs, risks, quality, procurement, communication, etc.

There is one more element that adds to the difficulties with change requests: the BA never has enough time or capacity to properly assess the change and its impact on the requirements and their traceability, and one of the most significant consequences is a limited ability to properly design the change, fully understand it, and, in turn, properly assess its impact.

One additional challenge change requests introduce is the *free-for-all* atmosphere: when change requests are accepted and become part of the project scope, there is no clear prioritization process around them, and we often simply go and perform that newly approved work with little to no regard to their relative priority or whether the work to build these features is truly immediate.

All in all, waterfall (deterministic) approaches can handle changes (through change requests), but they are not designed for high-churn environments. The change is triggering a significant effort (including costs and resources) and risk, in order to change what we had previously spent significant effort (costs and resources) to build—our project plan. It is reactive by nature—an afterthought and an attempt to backpedal from our previous incomplete, unclear, or insufficient efforts to define our project scope and plan around it.

So here come agile approaches that say, "Let's get a better handle on that change thing." How? First, we recognize that we do not know everything up front (and hence, we cannot finalize our plans early on and then simply follow them); and that things change on us (frequently and without notice) so we need to be more nimble about it. How nimble? We need to define short sprints (or iterations), lock in change throughout those iterations, and by doing that, allow the team to do what it was mandated to do at the start of the iteration. Based on the progress, emerging needs, the quality of the work we produce in the iteration, and any new ideas that may be introduced—all these new changes will be considered, prioritized, and performed in the next (or future) iteration.

We are now protecting the team from the constant interruptions that changes bring with them, and we are protecting the scope of the project from ideas that have not been thought through properly—from ideas that are less feasible to those that may appear urgent, but do not actually add sufficient value. Once we stop the work at the end of the iteration, we collect our thoughts and consider whether and how well each idea for change fits in overall. Does it protect the stakeholders, our investment, our commitments, the overall vision for the product, and the project's success?

This mechanism to handle change is part of a bigger system that agile provides which helps us produce stability and provide a sustainable performance over time—all changes that take place will be effectively priced based on their impact and their trade-off will always go back to the scope. Any additional item that is approved to take place will be included in lieu of a lower priority item that will now need to wait or even be bumped out. With a timebox, a set team size, a budget, and a set quality, all change impacts will revert back to scope trade-offs.

The big question that comes to mind is: what is so challenging about it? It sounds simple and straightforward. First, people get confused around the notion that no change is allowed throughout the iteration. It is easy to explain that we just freeze the mandated scope (backlog) of what we need to do for a short period of time (a couple of weeks in most cases), but in practice, some people complain that this is not really agile because we are all about change so why not allow change when we need it to take place? The answer is simple: if we allow change all the time, it brings us back to the reality (described previously) of the waterfall projects. In addition, we should all close our eyes for just a few seconds and envision the last time we had a senior stakeholder coming up with a new need. Now we need to think: how understanding were they about our request to wait with their idea until the end of the iteration? Not so much—it is not easy to follow a new set of rules and to be disciplined about how we do things.

Beyond people's confusion and adjustments, there is the trade-off part. We are programmed to think that trade-offs have wide-reaching impact and that when we add something to the scope there will be an impact on the budget, the schedule, and other elements. However, with agile, the impact remains confined to the scope: when a change request is accepted (new features are added to the backlog), something else, lower in priority, will pay the price. It will either be bumped into the future or removed altogether. But this is easier said than done; having the trade-off remain confined to the scope means that there has to be a prioritized list of features to build on the backlog and priority is synonymous with accountability. We all get the chills from even the thought of hearing: *everything is a top priority; everything is equally important.* Senior stakeholders struggle with accountability and with the ability to prioritize what needs to be done; it requires thinking, understanding what is at stake, considering other stakeholders' needs, and accounting for capacity, capabilities, and organizational context. It is much easier to simply tell the team to go and do everything, but agile environments are unforgiving—any problems or bad behaviors that are associated with lack of accountability quickly surface and impact the project in the short term.

Following a plan is about traveling within a safety zone. We know what we need to do, we know what the plan says, so we are just cruising in the safety zone. When things go wrong or do not go at all, we can always blame the plan (we didn't know), blame those who put the plan together (they didn't know), or blame reality (no one could have seen this coming). Coming out of the safety zone means that we put ourselves more at risk (not to mention accountable) and if things do not go well, fingers will be (and rightfully so) pointing at us.

Dealing with Organizational Change

One of the biggest and most damaging misconceptions about agile is that people think the organizational change component is simple, quick, and easy to handle. Agile practitioners try to communicate the message that it is not as simple as it looks, but this misconception continues to live on. The fast pace of events, the innovation it requires, and the disruptive element that agile practices have on most environments' *best practices* are going to face resistance because organizations cannot transform at the push of a button.

Before trying to handle the organizational change element as part of transforming into agile, we first need to understand the current state of our organization. Similar to any other type of change, or even a process improvement process, understanding the current state is the first step toward a positive change and success. Without a sense of what the organization values and how things get done, we will have a hard time finding the right way to do things moving forward. After all, when we go to the doctor asking for medication to relieve pain, the doctor first needs to know the cause of the pain since no one medication will be right for all situations.

The bridge to be built toward the future state of agile must address the organizational pillars and speed bumps, including existing processes, escalation procedures, decision-making mechanisms, work authorizations, reporting systems, governance structure, and the amount of rigor of these things. While it is applicable to review these areas for every agile transformation, larger organizations, government agencies, and organizations that operate in highly regulatory environments are more likely to require special focus on areas related to reporting, documentation, and governance, in particular.

Training, development, and certifications (in both agile as well as in organizational change) for team members and management are one common and important way to get the message across and achieve a certain level of consistency. However, training and certification are only one way, and while being a component, are definitely not sufficient to enable such change. Allowing external agile coaching to help facilitate the process, along with external help in organizational change, will help bridge the gap between theory and reality.

It is expected that any type of organizational change will trigger resistance in the organization. The resistance can surface for many reasons, but regardless of *why* there is resistance, it will cause issues and delays and it will have to be dealt with. The organizational change that agile adaptations introduce is complex and challenging, but is not outside of the range of other organizational changes for different reasons and triggers. The *Stealth Approach*[1] introduces small processes that move the organization toward agile methodologies without calling it that by name. This is an effective way to introduce agile-related concepts—through the back door—and potentially reduce resistance to the change.

The BA, Change, and the Retrospective

Business analysis skills are a natural fit for organizational change because BAs can help define the current state and produce a needs assessment to the decision makers in the organization. BAs can also help identify training needs and when things get rolling, BAs will provide the skills and experience along the way. Whether we need the BAs to help ensure that agile practices are ceremonies that are taking place as required depends on the team's discipline, but one way or another we need to make sure the team takes these agile events seriously. One of the most important mechanisms for process improvement is the retrospective.

Many people tend to take the retrospective with a sufficient level of gravity—they do not see the value in it. There is still some jaded sentiment back from the days of end-of-project lessons learned exercises, and they do not see how they personally add value to the process. With that said, we need to keep in mind that retrospective exercises are the engine of process improvement and they will achieve their potential only if we all actively participate and ensure that meaningful lessons are captured and applied.

For an effective process, we need to have the retrospective session right after the iteration review and demo, and split it into two. The first part should take place in a team-member-only format. It will allow the team members to open up and have a candid discussion about what happened; what has to change, start, or stop happening; and how we bring that to fruition. We should then move to the second part of the retrospective process where the product owner (or the sponsor) also attends. The team should now be able to present a more consolidated view of their own meeting's outcome with recommendations for the product owner to review and approve. It will also save time for the product owner as there is no need for him or her to attend both meetings. Finally, the BA will capture the action items, their goals, and their format, then along with the Scrum Master, will ensure that lessons are being incorporated into the next iteration and tracked for effectiveness. At the end of the upcoming iteration, the BA will present the findings of the lessons applied, along with a review of how effective the changes were and what side effects they had, if any.

If team members fail to contribute to the process, it will lack their point of view and the lessons will be incomplete, irrelevant, or even the wrong lessons to take. Some schools of thought call for a reverse order of the retrospective meetings; that is, starting with the product owner and subsequently having a team-only session, but this simply depends on the team dynamic, organizational needs, availability of the product owner, and the actual context on hand.

One more consideration in relation to a retrospective needs to be taken into account: the need for a meaningful retrospective session at the end of the project—on an organizational level. This is when senior management needs to look into the business case and the project's results and review what the original intent was compared to the project's outcome. This exercise can be very effective not only in looking at the project's performance, but also to take a candid look at the organizational needs, how they are articulated, and to what extent they have been achieved. It is important that the organization has a clear view of the impact agile introduces from a strategic point of view so that we do not get bogged down with the day-to-day practices and challenges related to agile.

APPROACH, READINESS, AND GOVERNANCE

This section is broken into three parts to help in choosing the right approach, assessing organization's readiness for managing agile projects, and looking at challenges and solutions in relation to agile and effective organizational governance.

Choosing the Right Approach

Agile is not the cure for all organizational or software development related challenges. It is important to ensure that we choose the right approach to managing our projects based on our actual needs and organizational circumstances, and not solely based on trends and on what others do. Many organizations decide to bring in agile approaches to their projects as a result of *peer pressure* (everyone else is doing it), instead of evaluating how suitable agile is for their needs and context.

Agile is applicable when projects take place in high-changing environments and/or when there is a need for early realization of benefits, and/or when there is a strong focus on achieving quality results (see Figure 2.1). For low-change projects where little to no changes are expected to take place throughout their life cycle, there is no need to establish all these mechanisms and efficiencies associated with agile (see Figure 2.2). Since agile approaches focus heavily on changing environments, being all set up for these conditions while they are not present will simply be a waste of time, money, and resources.

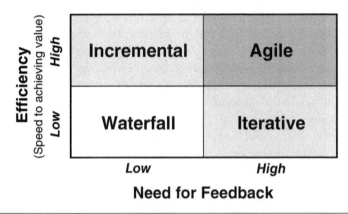

Figure 2.1   Agile applicability to *high*-change projects

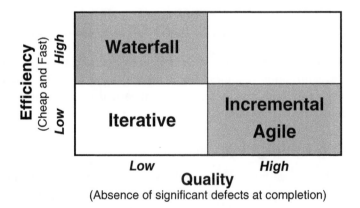

Figure 2.2   Agile applicability to *low*-change projects

Deterministic Approaches

The methods under consideration range from deterministic approaches that are plan-based to empirical approaches that are observation-based. Deterministic approaches (e.g., waterfall) focus on adhering to plans. The team spends an extensive effort identifying needs, defining requirements, making estimates, and finalizing (also known as baselining) a plan, and then following it. It is possible to incorporate changes into the process; however, these come at a hefty cost as there is a need to undo a portion of the effort that was taken to put together the plan in consideration of the new, changing needs. It introduces more risk to the project and inherently more cost due to the need to undo the plan, replan, and seek approvals for the new plan. Overall, for low-changing environments and when there is no need to seek early realization of value, waterfall methodologies are safe and simple.

Observation-Based Approaches

Observation-based approaches (e.g., agile) focus on delivering value to the customer and if the needs end up changing at any point in the project, we attempt to maximize the delivery of value to the customer by adapting to the changing environments. Planning is done in a different way: we start with a clear product vision, perform a high-level round of planning to come up with high-level requirements and rough estimates, and then from here the team goes through *spikes* of detailed planning and estimating on a rolling-wave-planning basis—one iteration at a time. Each time, the detailed planning takes place for the upcoming iteration, while focusing on what needs to take place next. The importance of focusing on what comes next is due to the need to maximize value to the customer. The latest set of information is the most relevant and typically represents the latest set of needs and performance levels—allowing the team to deliver value even if the value has changed since the previous iteration.

Iterative Approach

With the deterministic and empirical approaches at both ends of the same continuum, in between there are two more approaches: incremental and iterative, as seen in Figure 2.3. Iterative approaches focus on adapting to change and getting feedback. While there is only one delivery of value at the end (like a big bang delivery), iterative approaches are designed for changing environments and they are all about addressing the need for ongoing feedback; mini loops of *design-requirements-validation-build* with loops of feedback and fast-tracking design for the next cycle are taking place. Although there are multiple loops of

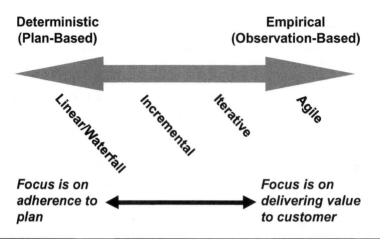

Figure 2.3   A range of methods from deterministic to empirical

value creation, there is no workable portion of the solution made for the customer until the very end, which, despite the effort for incorporating the feedback, may still lead to quality issues.

Incremental Approach

Incremental approaches focus on the delivery of incremental value at the end
of each iteration, hence allowing the customer early realization of value. Incremental approaches are not about adapting to change; while there is an ability
to adjust things moving forward, the adaptability to change is not intended for
what has already been delivered in the latest increment. It is strongly tailored
toward quality since the customer has multiple opportunities to see and even
use a working portion of the final solution, increasing the chance that quality
will be delivered. Incremental approaches, while having only a limited or even
secondary focus on feedback, include primarily four sets of benefits:

1. *Efficient reduction of scope*: the ability to reduce scope in later parts of
   the project in order to meet time/budget targets, while minimizing the
   reduction in value
2. *Increased quality*: multiple product increments at consistent intervals
   will produce ample opportunities for the customer to understand the
   direction and the value created—knowing about and seeing the path
   toward quality

3. *Reduced risk*: with multiple chances to see progress and to benefit from working portions of the final product, there is less chance for technical risk (less significant defects and technical debt remaining) and less risk of building the wrong thing (as the customer realizes pieces of the value throughout)

4. *Early realization of value*: the customer has the ability to use portions of the final product earlier, throughout the project, which helps the customer benefit from the product without having to wait until it is complete and deployed; and further, it reduces the overall cost of building the product because benefits can be realized throughout the project (see Figure 2.4)

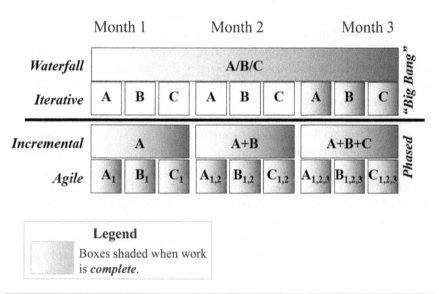

Figure 2.4   The agile approach is iterative *and* incremental

Back to Agile

Empirical approaches (agile), also known as iterative and incremental development (IID), compound the benefits of both iterative and incremental approaches. Agile approaches are designed in such a way that makes them open for feedback with relative ease to respond to change, and they provide all the benefits of incremental approaches, including incremental delivery of a working product, early realization of value, and overall lower risk. The focus on quality is also achieved through rounds of regression testing to help ensure that any product increment that is produced works with all previous increments.

Picking the Right Approach

When organizations fail to manage the projects under the most suitable approach, it will introduce challenges, delays, inefficiencies, and errors that could potentially compromise the project's quality and its ability to deliver value efficiently. The task of picking which approach is most suitable is an important one and should not be overlooked or done in haste. A lot is hinging on choosing the right approach for the current organizational needs and failing to do so may spell issues and problems—ranging from mere inefficiencies to project failures. The inefficiencies could be introduced as a result of, for example, picking an agile approach for a project in an environment that experiences little or no change. Doing so puts in place mechanisms and processes that are driven toward changing environments, but with little to no change, these mechanisms will go to waste. More severe issues, or even project failure, may occur when we select, for example, an incremental approach for a project that does not need early delivery of value.

Organizations should adapt an agile triage checklist in order to determine whether agile is the right approach for their projects, as seen in Figure 2.5. This set of questions will help reiterate the actual needs and challenges the project on hand actually has; and by that, help determine what the right approach should be. The process of going through the checklist is a process that is part of the organizational strategy—to ensure that the approach we take to managing projects is the most applicable and relevant for our needs. However, like many other enterprise or portfolio management activities, it is often the BA who leads the effort of identifying what the right approach to any specific project is—and further, ensuring all criteria, considerations, and stipulations are taken into account.

The BA is not the champion of the process to assess what the right approach should be—this should be mandated by a strategic branch or the project management office (PMO)—but the BA serves as an important part of the effort and it is common to find the BA leading this effort. With that in mind, the BA needs to understand the considerations so the due diligence process is focused, thorough, and above all, ends up pointing out the most suitable approach for the situation at hand. The BA's guide for choosing the right approach that is shown in Table 2.1. Depending on the decision-making process and on other organizational considerations, the BA may produce a report with recommendations on which approach to take, or alternately, produce a findings report, leaving the

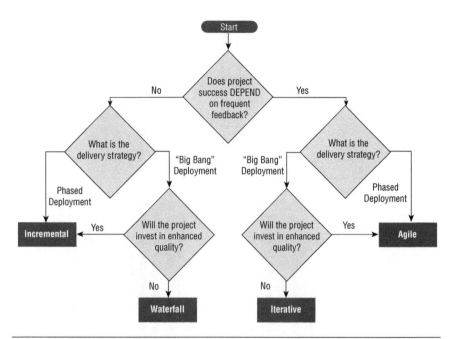

Figure 2.5   Choosing the right approach: agile is not the cure for everything. Follow the process flow: start by checking if there is a need for feedback and then follow the splits between low feedback (waterfall or incremental) versus high feedback options (iterative or agile).

decision making to the appropriate authority. Keep in mind that the recommendation the BA may make is still a recommendation since the BA does not have the decision-making authority here.

Pressure to Go Agile

Another common challenge that organizations face is the pressure to go agile. This is related to picking the right approach, since many organizations decide to adapt agile methodologies for the wrong reasons. Instead of doing agile projects in order to improve performance, they go agile due to *peer pressure*— because the competition does it or simply because they want to try something else (new) to improve performance. The results of such effort are very likely to lead to project failures and, in turn, these failures may lead organizations' leaders to blame agile for their failures. It is therefore important to go agile for the right reasons and only for projects that are suitable for agile approaches.

Table 2.1 Approach selection guide for the BA

Approach Characteristics	Deterministic (Waterfall)	Incremental	Iterative	Empirical (Agile)
Characteristics	Plan-based	Plan-based for each iteration, one iteration at a time	Observation-based, designed for changing environments with a high need for feedback	Observation-based. IID: Iterative and Incremental Development
Delivery of Value	Big Bang, only at the end	Delivery of incremental value at the end of each iteration, hence, allowing the customer early realization of value	Big Bang, only at the end	Delivery of incremental value at the end of each iteration. Producing a shippable product each iteration—allowing the customer early realization of value
Adaptability to Change	Very limited	Focus is not on adapting to change. Can adjust things moving forward, but the adaptability to change it is not intended for what has already been delivered in the latest increment	Designed for changing environments and is all about addressing the need for ongoing feedback. Mini loops of design-requirements-validation-build with loops of feedback and fast-tracking design for the next cycle	Designed for changing environments. Each iteration produces a workable, production-quality product that adheres to the last knowledge and information to maximize the delivery of value for the client

Focus on Quality	Limited focus on quality as it is not open for change and there is limited mechanism to adjust. Inherently, when the first chance the client has for feedback is at the very end, post deployment, it may lead to quality issues	Strong focus on quality; customer has multiple opportunities to see and even use a working portion of the final solution, increasing the chance that quality will be delivered. Regression testing helps ensure any product increment that is produced works with all previous increments	Although there are multiple loops of value creation, there is no workable portion of the solution made for the customer until the very end, which, despite the effort for incorporating the feedback, may still lead to quality issues	Strong focus on quality and value creation for the customer. Multiple rounds of creating product increments (working portion of the final product) and looping back to build the next increment based on the latest feedback. Regression testing helps ensure any product increment that is produced works with all previous increments
Benefits and Applicability	Simple and stable for low-change environments	Incremental (feedback is secondary) 1. Efficient reduction of scope 2. Increased quality 3. Reduced risk (technical risk and risk of building the wrong thing) 4. Early realization of value	Iteratively ensuring changing and emerging needs are addressed, but no early delivery of value	Compounds the benefits of both iterative and incremental approaches

Agile Readiness

Wanting to adapt to agile or even realizing the need to move in this direction does not mean that organizations are ready to manage agile projects. It is similar to other things in life: intent is not the same as readiness; however, unfortunately these two often get confused.

Beyond Approach and Beyond Intent

A big part of getting ready for an agile project is to identify what type of approach is most suitable for the organizational and project needs and contexts (deciding between waterfall, incremental, iterative, or agile). Knowing what type of approach is an important step because if we do not know what approach we need to pick, it is quite obvious that we will not be able to set the approach up, ensure alignment to organizational considerations, and properly position ourselves with the right processes, communication, reporting, and other mechanisms.

Even when we know which approach is most suitable for us, it does not mean we are ready for agile projects. We now need to assess our performance and practices for the purpose of knowing how far we are from being able to manage agile projects; and largely based on that, determine the steps, adjustments, and actions we need to take in order to make it happen.

What Is Readiness

Readiness is the state of being prepared for something; and the first step after identifying an approach to go with is to determine if we are ready or if we have what it takes to pursue the project. It gets easier if we have some sort of criteria or a benchmark to compare to and aim for. Agile methods focus on delivering value more efficiently to the business, and with a focus on increasing speed/throughput, agile methods tend to exacerbate existing people, processes, and systems issues in the organization. While agile methods offer many effective and efficient ways to achieve better results and deliver them faster, agile methods are unforgiving—when they face organizational issues, these issues tend to appear in a more acute and damaging form.

Agile Readiness Assessment

When making the effort to get ready for agile, there is a need to define an approach, set up criteria, and perform an analysis on how ready the organization is. The Agile Organizational Readiness Assessment™ is an effective approach for checking an organization's level of readiness to take on agile projects.[2] It is an assessment that covers several areas that are important for agile success. The assessment is not a form of audit as to how things are in the organization, but rather it is a more advanced form of assessment: it checks how current and existing processes and practices in the organization are aligned with and suitable for agile. It checks whether the way the organization currently conducts business is conducive for achieving success through agile projects.

Performing the agile readiness assessment often produces a positive type of *side effect* that further benefits the organization. While the assessment provides a reading into how prepared the organization is to manage agile projects, it also points out a series of organizational issues, problems, and challenges that are likely to be in the way of a successful agile project. These are organizational issues that are not directly related to agile, but despite that, they will serve as hurdles on the way to achieving agile project success. This helps organizations realize *economies of scope* benefits that agile introduces: although agile is unforgiving in the sense that projects will stumble on organizational challenges that appear to be unrelated to agile project management, agile also helps to *point out* and in turn *fix* other organizational issues.

In fact, even if the organization ends up not proceeding to manage agile projects, the Agile Readiness Assessment can help the organization improve its processes regardless of agile, and with it, improve its bottom line. This is because agile projects *suffer* from any type of organizational issues. It is important to note here that no matter how ready organizations are for agile projects, it is not expected that an organization will transition all of its projects to agile. As explained in the *Choosing the Right Approach* section, agile is not applicable to all environments, all projects, or all circumstances. It is expected that an organization will have different situations and groups, some of which will not require the benefits and features that agile projects offer. Agile (which is an incremental and iterative delivery) will benefit projects that face significant changes and that require early realization of benefits. Waterfall, on the other hand, is suitable for projects where little to no change takes place and where the customer does not need early delivery of value in increments.

Performing the Assessment

The assessment identifies seven areas to measure that are related to agile principles, best practices, and concepts; and it checks how close the current organizational processes and practices are to being ready for agile projects, how aligned they are with agile principles, and how critical they are to success. There are seven items under consideration for the assessment that closely represent practices that need to be in place for delivering success through agile projects (see Figure 2.6). These seven items are a result of extensive analysis, observations, and strong experience from a wide range of agile projects and organizations over many years. The breakdown of the tools and techniques to assess and

Criterion	Score
IID Practices	4
Evolutionary Design	3
Planning & Estimating	5
Communication	6
Collaboration	5
Business Value Focus	3
Metrics,Tracking, & Forecasting	6
Avg	4.6

Agile Health: Organizational Readiness Assessment Sample

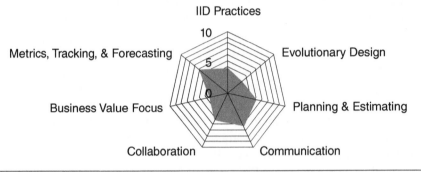

Figure 2.6   Agile organizational readiness assessment

measure them are proprietary. The readiness assessment focuses on checking current and existing practices, processes, tools, and actions in an organization.

The assessment begins with a general, high-level scan of the health of the delivery organization as a whole; and while it is not an exhaustive review, it typically identifies issues that can be addressed quickly to improve the success of an agile adoption. Through interviews, conversations, observations, and documents review, the assessors collect the information and analyze it against the tools' parameters to determine a score for each of the seven criteria under consideration:

1. *IID Practices*: A review of how current-state processes and practices align with agile concepts. Iterative development is about how a team is set up to handle changing environments (the intent here is not focused on existing change control mechanisms) and incremental development, which is about the team's ability to deliver increments, or *chunks* of working products, and hence value to the customer on an interim and ongoing basis.

2. *Evolutionary design*: Agile, being an empirical approach, is all about evolutionary design. It refers to the way the product is designed to allow incremental product releases on a frequent basis and dynamic plans that are simple to implement, allowing for changes and adjustments.

3. *Planning and estimating*: Looking at planning and estimating practices to determine how accurate, consistent, and realistic the estimates are and the team's ability to deliver on the plan's objectives, while welcoming necessary changes in direction along the way.

4. *Communication*: Agile methods require effective, efficient, and streamlined communication. Whether the team is colocated or not, this item evaluates the quality of communication as it pertains to agile needs—including flow, style, consistent use of terminology, formality, escalations, and decision-making processes.

5. *Collaboration*: While communication and collaboration are related to each other, they are not the same. There may be situations where an organization's score on these two items is not consistent, where one item gets a significantly higher or lower score than the other. It often surprises stakeholders who view these areas as related and therefore correlated. While communication refers to the flow of information and the mechanisms to enable it, collaboration refers to how effectively team members work with each other and how different areas of the organization work with one another. It refers to the project's ability to work effectively with functional areas, subject matter experts, other projects, resource managers, and senior management. Naturally, it also refers to collaboration within the team. It is common to see environments where information flows even though they lack collaboration, as well as the opposite—strong collaboration despite broken communication mechanisms.

6. *Business value focus*: Many organizations make a mistake by optimizing flow and efficiency, while failing to focus on business value. The teams that they create may be efficient, but they may be building the wrong thing. By selecting the right approach for the situation, it will be easier for organizations to add the right support systems (i.e., tools, techniques, processes, and approaches) to ensure that everyone focuses on delivering business value. Focusing on delivery of business value is something that needs to be present in every organization, regardless of whether they use agile, waterfall, or any other approach. Failing to focus on business value will serve as a major hurdle when attempting to transition to agile.

7. *Metrics, tracking, and forecasting:* Reporting mechanisms need to be in place, ensuring focus on business value and on value-added activities. We live in an era where we are inundated with data, but many organizations fail to translate this data to meaningful information that can serve as decision support. With the help of BAs, it is important to have the ability to take all the relevant data and extrapolate, analyze, correlate, integrate, and put it in context so that decision-making stakeholders have the ability to make timely, informed, and correct decisions. When organizations struggle with measuring, tracking, and reporting about their projects, it is a bad sign for any type of environment, but it is an alarming sign for anyone who even thinks about going agile. The reason for the elevated concern when going agile is due to the fast pace, the ability to work effectively under ambiguity, and the strong need to come up with consistent estimates. Problems in this category are almost a showstopper for a newly formed agile team and adjustments need to be made to increase the chance for success.

A Low Score Is Not Necessarily an Alarming Sign

It is expected that most organizations will demonstrate low scores for most items. A low overall score is an indication that the organization is not prepared for agile projects. However, the reason most firms are expected to have low scores is quite simple: the organizational readiness assessment is performed mostly in organizations that do not practice agile or in ones that have failed to realize agile benefits and are looking for the reasons why it happened.

Note that when stakeholders view the low scores, it tends to lead to resistance because they view the assessment as criticism on the way things get done in the organization. Stakeholders tend to default to viewing the assessment as an audit, and they tend to interpret the findings in a personal manner that implies that they are underperforming. It is important to properly present the findings and to set up the review in the right context to ensure that the assessment's findings are viewed for what they are: a snapshot to determine how prepared the organization is for managing agile projects with circumstances, practices, and processes as they are.

Once stakeholders are on board, it is time to provide recommendations based on the scores of the individual items assessed. The recommendations involve tangible action items, order of priority, timelines, and sequencing—so the right action is taking place to ensure stronger alignment to agile principles. The recommendations report further increases the buy-in to the process among stakeholders because it now clarifies that even a low score does not mean that the organization is ill-prepared to manage agile projects. The action items help stakeholders realize that the lower score is an indication of the current state. In most cases, the desired state is not difficult to achieve and a series of small steps will usually suffice. Therefore, it is safe to say that a low score in the readiness assessment is not necessarily a major cause for concern. The important thing about the score is to ensure that it is truly reflective of the conditions in the organization so that necessary action and adjustments can be made to improve the alignment and readiness to run agile projects.

Assessment Bottom Line

Overall, the agile organizational readiness assessment does not serve as an indication or make any implications about current performance. It simply shows the organization which areas the stakeholders need to focus on to ensure that the organization is ready for agile projects. The recommendations derived from the assessment serve as a guide as to which areas to focus on in order to become more prepared for agile projects.

Agile Governance

There are only three roles in Scrum projects: product owner, Scrum Master, and team member. The *fathers* of Scrum refused to use the term PM to avoid any type of association with the traditional role of a PM, which is often policing the project team, being consumed in putting out fires, and *herding cats*. The Scrum reference to the Scrum Master is of a servant leader who is there to protect the team and facilitate the agile process. When referring to the product owner, the nature of the role implies more than just leadership and governance. The product owner is a true sponsor with timely, frequent, and meaningful involvement throughout. When the product owner cannot provide the required level of involvement and support that the project needs, it may lead to a failure to achieve project goals. With that said, an effectively utilized BA can serve as the eyes and ears of a product owner and help the project make progress even with a lack of sufficient involvement by the product owner. The intent of the BA's role here is not to replace or override the product owner, but rather to work within the confines of the mandate provided by the product owner, consolidating inquiries and feedback, and providing context for everyone involved.

The BA's role in an agile project can provide support to the product owner, the team (requirements, prioritization, testing), and the Scrum Master (facilitation, process, escalations, product related items).

Agile and Governance

Most large organizations are governed by strict processes that are designed to protect the shareholders by ensuring that funds are invested properly, risk is considered and minimized, projects align with organizational strategy, and the organization has sufficient capacity to handle new challenges and opportunities. While excessive governance can hinder the ability to deliver business value efficiently, a lack of governance leads to chaos and waste. There is a need for the business leaders to ensure a balancing act between control and efficiency. In many organizations, especially large ones, as well as government agencies, the balancing act tilts heavily toward control—at times excessively. It materializes through adding red tape, having multiple layers of approval, riddling processes with duplication of effort and redundancies—all leading to slowdowns, major waste, and at times, untimely, late, misinformed, and incorrect decisions, or even a complete failure to act.

Many believe that agile and governance do not mix. Some of the common misconceptions about agile refer to it as some sort of a *free-for-all* with no planning, no documentation, and no governance. Agile methods evolved to combat excessive governance on high-change projects, but many still believe that agile methods cannot succeed under strict governance models. What may be a surprise for many is that in the real world, agile methods cannot be used in large organizations without adhering to standard governance models.

Standard Governance Model

Surprisingly, agile methods are not very different from waterfall or linear processes in how they are governed. A standard corporate governance model, as illustrated in Figure 2.7, has five governed milestones throughout the life cycle—from having an idea until realizing the value. These five steps are governed with an organizational gating process, referring to the full life cycle of an idea all the way to value realization within which there will be a project. The details of the five steps are shown here. After each step, there is a gate where decision makers will need to determine whether or not to proceed to the next step.

1. *Concept*: This is when an idea is introduced along with its possible benefits. If the decision makers in the organization see the value in the ideas, it passes Gate 1 on the way to an analysis process. In certain cases, seed funding may be allocated at this point so the organization can conduct the next step in the process.
2. *Opportunity Analysis*: The second stage is an opportunity, or marketing, analysis. This process checks whether the idea or concept will actually be accepted by the customer. This step involves some good old business analysis work for information gathering—including interviews, focus

groups, and other measures to check if and how much the customer will like the new idea. This step leads to Gate 2, where a decision has to be made on whether to proceed or not.

3. *Preliminary Business Case*: Once we pass Gate 2, we proceed to conduct a preliminary business case to check the justification of the idea. We now believe that it is a good idea (Gate 1) and that it is likely to be in demand by our customer(s), and it is time to check whether there is justification in continuing to pursue the idea. The business case study is supported by a feasibility study that goes beyond justification, and it checks whether the organization has what it takes to continue pursuing this initiative. Once again, business analysis activities are required, including performing high-level solution design and requirements, high-level estimates and timelines, and risk analysis—so there is a read on the cost-benefit analysis and the potential return on the investment. If the idea is deemed financially sound *and* feasible, it passes Gate 3.

4. *Detailed Business Case, Pilot, or Proof of Concept*: At this stage, we validate and commit to design and approach; validate the high-level estimates from the previous step and commit to revised, more accurate estimates and timeline; take measures to reduce risk; and baseline the plan. We now have a project—and this leads us to the last corporate review of this initiative, which is Gate 4.

5. *Full Execution of Deployment to Realize the Business Case*: After passing Gate 4, we move forward to building the solution, deploying it, and realizing the value of the business case. This is the last step in the process, and after Step 5 there is no gate since the project and the idea's life cycle both come to the same end—realization of value.

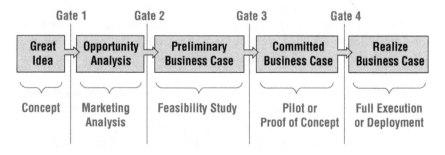

Figure 2.7   Standard corporate governance model

It is expected that the chances of passing a gate improves with time, which means that the later the gate, the greater the chances of passing the gate. With that said, these gates still serve as checks and balances to ensure that at any point we do not proceed with the initiative unless it is the right thing for the organization. Additional characteristics include a consistent increase in the investment of time and money throughout the initiative's life cycle, as well as a declining trend on the amount of risk, especially after passing Gate 3.

These five steps describe a standard governance process that takes place in most organizations. The level of formality changes from one organization to another—and so does the level of rigor and formality. Organizations that fail to have an effective and enforceable gating process may find themselves with the wrong initiatives and bad ideas that should not be implemented, but proceed to become projects that are then deployed due to the lack of proper ability of the person in charge to put his or her foot down at certain intervals and stop the idea. Other organizations may suffer from a different type of problem, where too much rigor and heavy-handed gating processes kill good ideas in some form of analysis-paralysis.

If the governance model is dysfunctional or broken in a waterfall/linear setting, it will fail any attempt at agile.

Aligning Agile Methods into the Standard Governance Model

We already established that it is highly unlikely that organizations will completely *convert* to agile, having no waterfall projects at all. Most organizations will have some sort of a combination of projects, some agile and others linear—based on their circumstances. In line with that, there will need to be two governance structures: one that is set for linear projects and the other for agile projects. The good news is that there is not much difference between the two governance models and that they can both coexist and each serve for the respective types of projects.

The task of aligning agile methods to the standard governance model is not complicated. Of the five-step process, the main difference will take place in Step 3 (Preliminary Business Case, leading to Gate 3) and in Step 4 (Detailed Business Case, Pilot, or Proof of Concept, leading to Gate 4). Otherwise, the governance models for agile and for waterfall projects are similar and follow the same logical flow.

During Step 3, waterfall projects go through a detailed design and estimating process; and it is relatively costly and time consuming to perform all of these planning activities up front. In contrast, agile projects will go through a similar set of steps and gates leading to this point, but will have a different Step 3 where only high-level design and estimates will take place. These activities consume a relatively small cost and take a short time to plan up front. These activities are usually part of what is called *iteration zero* in agile projects.

Once the initiative passes Gate 3, Step 4 will once again have differences between agile and waterfall approaches. Waterfall models may conduct a pilot or go directly to full execution, depending upon the risk level. However, agile models usually perform a short pilot to reduce risk. After 2–3 iterations, actual performance (*velocity*) is identified and can be used to forecast completion with reasonable accuracy, similar to earned value. These are followed by completion of detailed design in waves and during each iteration—following the principles of rolling wave planning.[3]

After passing Gate 4, both agile and waterfall approaches continue with the project until it is complete. This corporate or organizational governance model is not to be confused with project governance. As demonstrated in Figure 2.8, the governance models are similar for waterfall and for agile (with the differences specified in this section); however, the projects under each methodology will be managed differently according to each respective approach. Waterfall projects will conduct one robust planning process at the start, followed by an extensive process to create value, change, test, and fix defects. Agile projects will be broken into iterations, or sprints, where during each iteration the team will go through a round of detailed planning, building, and testing to produce a production-quality product. The team will proceed at the end of the iteration, upon acceptance of the product, to the next iteration that represents a similar cycle to produce the next product increment.

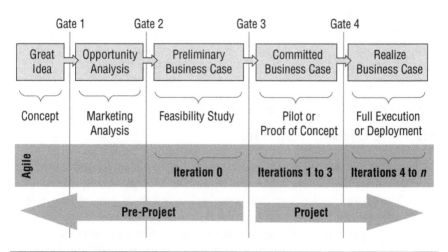

Figure 2.8   Governance models alignment summary

Too Much Focus on the Process

Agile introduces new and different ways to deliver value in projects. Agile involves new processes and requires people to behave differently and respond to situations in different ways. Naturally, it involves organizational change and, in turn, the change takes people outside of their comfort zone. When people face new situations, uncertainties, and fast-paced environments, and when people are required to move forward despite the ambiguous conditions around them, they tend to cling to whatever is familiar. They become conservers (i.e., trying to hang on to existing processes) and they resist the change toward new processes. These behaviors create yet another side effect: people also become more focused on the mechanics and their creativity and innovation get diminished or even disappear.

Many new teams that embark on agile projects for the first time suffer from these afflictions. Instead of maintaining an open mind, looking around for opportunities, keeping an eye out for waste, helping each other, finding creative solutions for problems, and innovating new and more efficient ways to produce value, they shut down into their little box of familiarity that protects them and stay there. Their reduced capacity prevents them from paying attention to the things around them and to potential efficiencies; and instead, their focus on the processes, ceremonies, and new ways of doing things essentially consumes them and their capacity. These teams often end up focusing too much on being agile and not enough on doing the work; failing to realize that the main goal of agile is to clear impediments and produce value. When agile team members (including the Scrum Master and even the product owner) lack experience and when there is a heavier focus on the process, rather than on the work products, it might lead to issues. A clear breakdown of the BA's responsibilities and any meaningful support the BA can provide to the team, the Scrum Master, and the product owner will help stir the team toward stronger performance.

RESOURCE MANAGEMENT AND THE MATRIX ORGANIZATION

Agile methodologies call for dedicated teams, but in reality relatively few organizations allow that. There is an expectation that once an organization decides to go through agile adaptation, the organization will magically transform from the tough reality of the matrix organization to a system where resources are dedicated to the agile team—or at least will report to the project exactly when they need to. If the organization manages to transform itself and deliver on resource allocation and commitments, it will be easier for the agile project to deliver on the agile promise. However, in many organizations we see challenges with resource allocation: resources not reporting to projects on time, failing to stay on the project for the time needed, or being replaced by resources with skills or experience that are less relevant than the project's context calls for.

While agile projects are unforgiving, and the pain of resource allocation issues will be felt immediately and will significantly compromise the project team's ability to deliver on the iteration's goal, the team will be able to identify the issue immediately and quickly *price* the impact of the resource problem on the project. Being able to articulate the true impact to stakeholders, and most importantly to the product owner, improves the project team's position and everyone's ability to take action to reduce the impact of these issues.

In a waterfall project, the team will have a hard time identifying the impact of resource allocation issues and the mentality that *we can make up for it later* may prevail, further reducing the ability to truly identify the impact on the project. By the time the project's performance is actually impacted, it is too late to act on it or to find ways to address the issue effectively, not to mention proactively. In this sense, waterfall projects (while not to be blamed for such issues) are a good environment to hide people's bad behaviors and other issues, such as resource allocation deficiencies. Due to the long timelines, cycles, and planning horizons, it is difficult to identify the impact of issues, or even the existence of issues, and it is easy to believe that we will be able to fix them downstream. In reality, by the time we reach the point where we can fix the situations, too many issues are piling up, and their impact on the project's performance and its ability to deliver on the objectives is too severe.

Cross-Project Coordination and Dependencies

If an organization cannot guarantee dedicated resources for agile projects, it is important to establish a mechanism that although it will not cancel the negative impact of the resource allocation issues, it will at least minimize it. Cross-project resource coordination and the concept behind it are inspired by the concepts of critical chain project management (CCPM). CCPM is a methodology for planning, executing, and managing projects in single and multi-project environments.[4] The main principles behind it are to identify critical resources (known as drum resources)—these are resources that are shared by multiple projects and that are in high demand—and determine how to ensure that we utilize them to their maximum availability (not over-allocating them but ensuring that they are not idle at any point).

The technique to coordinate is quite simple, but requires genuine collaboration across projects and with the functional area managers who *own* the resources. Without entering too much into the concepts behind CCPM, the technique to streamline the availability of the drum resources is by placing strategically located buffers that will protect the availability and allocation of the critical resources, once they are scheduled to perform work in our project. To illustrate, let's assume that a resource is scheduled to report to my project on a Wednesday for a two-day-long task. To ensure that this resource can perform their work, our project should finish all predecessor tasks related to this resource's work by the end of day on Monday, leaving Tuesday as a buffer (known as a feeding buffer). This buffer may appear to be an inefficiency; yet by using this buffer, we can better control the inefficiency, and ensure it is contained to the buffer. The size of this buffer will be determined based on the risk in the tasks preceding it, and the risk of what might happen if the drum resource will not be able to complete their work.

The mechanism to ensure cross-project collaboration starts with identifying which projects are likely to compete for the same resources for the upcoming period (or iteration). The intent is to coordinate the resources in intervals that are no longer than two weeks since the longer the planning horizon gets, the more issues we are going to see come to the surface due to the lack of ability to provide exact plans for resources and tasks for the long term. The analysis should be done by a PMO or by the BAs who work on the respective projects. This analysis will identify potential resource conflicts by different projects.

The next step, once there is an understanding of potential resource conflicts, is to conduct weekly or biweekly meetings (depending on the project's velocity and the length of the iterations). The attendees should be the PMs from all of the projects involved, the functional managers who *own* the resources, and, if needed, the BAs involved as well as the corresponding project sponsors and product owners. In the meeting, the participants will assess the situation by reviewing the following factors:

1. *The priority of each project*: Priority is an organizational matter and it is determined by senior management—sometimes in conjunction with strategic needs. High priority projects are not necessarily going to get the *right of way* here solely based on their priority since there is also a need to check their relative urgency.
2. *The urgency of each project*: Measuring urgency is a difficult task since it is subjective and changes within days or even hours. Due to the subjective nature of the term urgency, there is a need to make it less nebulous and more measurable so we can compare apples to apples. There are three criteria to measure urgency, listed here in no particular order, as they all

need to be assessed: (a) the severity of the situation—considering organizational, project, and customer objectives. This includes dependencies on whatever is at stake, the impact of the situation, and the risk of not completing the work; (b) the timelines and deadlines associated with the project and the tasks on hand; and (c) additional considerations—including enterprise and environmental factors, context, and the stakeholders involved.

3. *Each project's resource needs*: It is necessary to assess the criticality of these resources to the situation at hand and when the drum resources are going to be in demand by the project, including for how long.

4. *The status of each project*: This is not a straightforward decision since at times we will allow the project that is doing ok to proceed; while on other occasions, there will be a need to advance a project that is not performing well.

The result of the meeting is going to be a timetable that maximizes the intended allocation of the drum resource(s) and a workable plan to maximize the benefits for the organization by promoting the most important (and urgent) projects according to the latest set of information available. It is not a perfect plan, but it is doable in the sense that it provides a realistic resource allocation to the projects involved that is based on resources' actual availability. If additional issues emerge throughout the following two weeks, at least our starting point is realistic and, at that point, we will need to deal with the situations as they emerge. With a lack of such cross-project coordination, the projects are all scheduling themselves in silos, expecting resources to report to the projects whenever they are scheduled. However, with resources being over-allocated, all of the projects that experience the over-allocation of resources will be negatively impacted. The impact will be hard to measure because at that point we become reactive, stressed, and even panicked in dealing with the situation—and we are likely to make misinformed, rushed, and potentially incorrect decisions as to where to direct resources.

With the proposed collaboration mechanisms, the projects will stagger their progress around the availability of the drum resource by placing feeding buffers that are strategically located to absorb risks and minor schedule fluctuations with the intent to protect the drum's availability. The mechanism for cross-project collaboration gives us a realistic plan to work with and it puts us in a proactive position to deal with potential situations before they hit us.

This approach is also very effective in environments where there is a combination of agile and waterfall projects. Such environments are more prone to issues around resources getting *stuck* in the waterfall project, jeopardizing the performance of the agile project that subsequently depends on the resource coming from the waterfall project.

The cross-project collaboration is dependent upon realistic and accurate estimates and identification of resource needs. It is expected, especially in agile projects, that the team will be able to provide detailed task-level planning for the upcoming iteration. There is also a need for projects to report any type of issues in relation to the drum resources as soon as possible when these issues surface to maximize the amount of time we have—and with it the options that exist—in order to minimize the negative impact of these resource-related issues.

The BA can serve as an important source of expertise for the estimates, communication, coordination, and candid and proactive approach that is required to make this cross-project coordination effort work.

DAY-TO-DAY CHALLENGES

Beyond the impact of the organizational change discussed earlier in this chapter, there is also going to be an impact on the individuals within and around the agile project. There is a series of new conditions and new ways of doing things, and they will trigger discomfort, anxiety, stress, and uncertainty. When first working in agile projects, many people feel outside of their element since they are being taken out of their comfort zone. In the short term—but hopefully not for the long run—the performance of team members may be reduced due to the myriad of new and different ways of doing things.

Working in an agile project means that some fears and concerns may materialize for some people—especially those who had misconceptions about agile projects. It is expected to see most of the following items interfering with team members' work:

- Fast pace, daily builds
- Increased and earlier involvement by testers
- New ways of estimating
- Different techniques for collaboration within the team and across projects
- New reporting mechanisms and information radiators
- Different and more frequent interactions with stakeholders, the product owner, and the customer
- The *war room*, which is a dynamic, and at times, hectic and often noisy environment for the team to work together, collaborate, and engage each other

- Frequent checks and balances, including a daily check with the Scrum Master or PM, a daily build, a stand-up *Scrum* meeting to start the day, and constant interaction all around
- Different reporting mechanisms (information radiators), including a status and progress report to be updated daily in the form of Kanban boards, burndown charts, and task boards
- Expectations that resources, for the most part, are more well rounded and can perform a broader variety of roles
- A different way of treating the project success criteria. The traditional way to define success through scope-time-cost-quality, as part of the competing demands, is managed differently through agile. While waterfall projects may change any one of these variables when there are problems or issues throughout the project, agile methodologies *lock-in* time (defined iteration durations through time-boxing), cost (with defined durations; and smaller, defined teams to better estimate project costs), and quality (with every iteration we deliver a *shippable*, or production-quality, product that can potentially be released to the client). This means that scope is the one variable that will change (or rather be reduced) because we will *bump* out lower priority features when the project cannot deliver on the full scope, while still maintaining the value creation for the customer.
- Change control of the project scope is done in a fundamentally different way. Since we *lock-in* the scope at the start of each iteration, no changes are allowed except for an efficient reduction in scope by removing lower priority items if we must in order to allow progress with higher priority user stories or features. As new needs, changes, defects, and other considerations emerge during and at the end of the iteration, along with the feedback provided by the product owner and other stakeholders in the demo at the end of the iteration, changes will take place. But the changes will be submitted as one *omnibus* change (to make it easier to manage and to reduce paperwork and increase efficiency). The product owner, along with the team, will go through the backlog maintenance task and determine which stories and features will be included in the upcoming iterations and at what trade-off (i.e., which stories will be removed from the iteration in lieu of the changes and what will happen with those stories). This is a different way to handle change to the project and it may trigger some challenges in adjusting to it.
- Many people have a hard time dealing with ambiguity since there is naturally a strong need to seek more information regarding situations they

face; but in agile projects we need to learn to live with ambiguity. In fact, we should encourage ambiguity until further information is needed. Our plans and estimates should be just *good enough*, which means that we should not pursue more detailed information about requirements or estimates until it is time to build them in the upcoming iteration. The need to perform under ambiguous conditions takes a toll; and many people struggle getting used to this way of doing things.

- There will be controversy around whether there is a need for a BA. Scrum and other agile methodologies do not call for the need to have a BA on the team. In fact, there is an expectation that team members will bring BA skills to the table with them and demonstrate business analysis activities as required. This may create tension in environments where there are BAs who may abruptly find their roles being questioned. In agile projects where there are no BAs, it may lead to the need to adjust the way things are done. While it is expected that team members demonstrate BA skills, there might be challenges around it. We should also keep in mind that when people need to wear two hats and perform two different roles, it is common that one role is more dominant (i.e., a developer who needs to also perform as a BA may be naturally drawn to the development work, pushing the BA tasks out and at times failing to perform them). With the *secondary* role being pushed aside, the project might suffer from the lack of timely or sufficient BA skills.
- Prioritization is important for any type of project setting; however, it also implies accountability. In non-agile environments, it is common to have situations where tasks and requirements are not properly or sufficiently prioritized. In a way, when a senior stakeholder tells someone to do something without prioritizing it in relation to other things on the docket, it is a form of passing (shirking) responsibility. In agile projects, we cannot afford the *luxury* of having user stories or requirements unprioritized.
- There is a growing need to automate testing in order to accommodate its growing role in agile projects. Chapter 8 of this book deals with the role of testers and testing in agile projects.

The *Diluting Effect*

In the earlier days of agile, the organizations that managed agile projects were relatively few and far between. However, agile has been and is increasingly becoming the *approach of choice* with a dramatic increase in the number of organizations that manage agile projects, in the number of overall projects that are managed by agile approaches, and in the number of agile practitioners.

While agile methodologies evolve and improve, with more and more agile projects taking place, there are some dangers that may cast a shadow on the ability of organizations to fully realize the benefits from agile projects.

Organization Type

It starts with the types of organizations that *go agile*. In the past, agile projects took place mostly in small technology companies, which by nature are fast-moving, flexible, and nimble with quick decision-making processes and small, integrated teams with visibility and open communication across all levels of the organization. The DNA of these small organizations was strongly aligned with agile principles, and the organizations had the ability to do things fast, make low-cost attempts for innovation, and quickly stand on their feet with minimal damage or losses in the event of a failure.

There have been many attempts to scale agile within larger organizations from an increasing number of industries attempting to transition into agile projects. While it is clear that agile principles are applicable to many industries beyond information technology, in some industries agile methodologies are not very mature. Scaling agile is not only about moving across industries, but also about managing larger agile projects with larger teams. Agile is mostly effective with small teams of up to 12 people (and some will say that the number is lower), but it can be scaled. An example of such scale is a Scrum of Scrums, where the teams are kept small, but they are all coordinated and work together toward their common goal.

There are many success stories regarding attempting agile in *new* industries. However, it does not always go smoothly when more and more *traditional* or *conservative* industries attempt it. Large financial institutions that are typically considered more conservative press forward with agile showing varying levels of success. Even government agencies that are considered conservative, slower moving, and generally less adaptive to disruptive changes move toward managing agile projects. When large banks or government agencies attempt agile, it introduces a large-scale challenge in the form of organizational change. Agile adaptations in such organizations typically involve more resistance and an increased need to adjust processes, policies, and procedures to accommodate the agile way of doing things. These changes, in turn, produce even more resistance since they involve changes in governance processes and significant adjustments to the way PMOs operate, along with changes to reporting, decision making, best practices, scope changes, prioritization, communication, and escalation procedures. For many organizations, these changes—along with the changes to the roles of individual team members, the PM (or the Scrum Master), the sponsor (or product owner), and the increasing role of testing—all trigger resistance as people are brought outside of their comfort zones.

Team Members

In the past, it was easy to handpick top performing resources to become part of the agile team. There was strong competition to enter the exclusive *club* of employees who were working on agile projects and it was easy to turn the agile principle into a successful reality of high-performing teams. However, with the rapid growth in the number of agile projects and in the number of organizations transitioning to agile, more and more teams have become agile. This may lead to a *dilution effect*, where not only the strong performers are assigned to agile projects, but we now assign resources that are not necessarily our top candidates for the job. The intent here is not to offend anyone, and this does not mean to imply that these resources are not good people with strong skills, a wealth of experience, and excellent backgrounds. However, not all people are suitable for the style and the pace of agile projects and not everyone has the right style, communication skills, and ability to collaborate in a way that will be conducive to the agile project.

As a result of going *down the list* of qualified resources, an increasing number of resources being assigned to agile projects are not fully suitable to work in these environments; that is—they are not the ideal candidates. While many of the agile principles and ceremonies are trainable and can be acquired, some keys for agile success are more difficult to acquire and to adjust to. It is clear that no two people are the same and, therefore, not all resources can fulfill all roles, nor can they all be successful in agile project environments. This *dilution* means that with less than ideal candidates on agile projects, we are less likely to deliver success. Further, as we continue to go down the list of candidates, and as more and more projects become agile, we have an increasing need for more resources to fill the roles in these projects. It is therefore expected that we are going to face more challenges with agile projects in the future.

Another thing to consider is that people tend to revert back to old, familiar behaviors. We need to keep in mind that it is not the waterfall methodologies that have caused the problems in product development as we know it, but rather, the practices, behaviors, and tendencies of people are what led to these problems. We already established earlier in this chapter that waterfall is not a problem in and of itself, but it is an approach that harbors bad behaviors and allows people to procrastinate and cut corners. It is the people who conduct themselves in a problematic fashion that leads to project problems; and putting people in agile environments will not automatically transform them into conforming, process-adhering individuals. Agile has multiple checks and balances, along with mechanisms that allow controls to more or less micromanage the team, without making it look like we do so. The mechanisms that agile offers are highly effective; however, when people do not follow the agile process, fail to participate in agile events, and refuse to report on their actions and performance, the agile project will start faltering to the degree that it will no longer offer the promises and the benefits that agile offers—until, ultimately, the product and the promised benefits fail.

The keys to continue maintaining the cohesiveness of agile teams is long:

- Training
- Coaching
- Education
- Effective leadership
- Strong project sponsorship
- Stellar communication skills
- Transparency
- Trust
- Hand-holding
- Improving
- Challenging and adhering to processes
- Discipline
- Eye for waste
- Efficiencies
- Continuous improvement
- Evolving best-practices

However, even these will not be enough; there is also a need to ensure that only suitable projects go agile, that there is support in the organization for agile projects, and that there is support within the team. The latter—support within the team—can be achieved with the installment of an effective agile BA to complement team members' skills, knowledge, and experience and to overcome challenges, deficiencies, and tendencies. All in all, the agile BA is crucial to ensuring there is someone in the right place and at the right time to cover for whatever any team member may fail to deliver.

The agile team can benefit tremendously from a BA. In agile our performance is measured as a team, and it is much more important than individual performance. However, the team is comprised of individuals and we need to ensure that all individuals do what they need to, when they need to do it, and to the extent that they need to. This task is not easy—the PM or Scrum Master can only partially do it, thus the BA is the perfect role for ensuring the team's performance is optimized and maximized. Originally, the intent behind creating the role of the BA was the success of the team to deliver on project goals.

In fact, the BA can hardly be viewed as a cost center since the BA's value is not measured individually; the value that the BA produces is mainly for the benefit of the team—and this is exactly what is needed in agile projects: ensuring that the team works effectively *as a team*, rather than as a collection of individuals.

Bias

This challenge further reiterates the need for having a resource with the title of BA in agile projects. When business analysis skills are shared by multiple team members, there is the danger we previously discussed: that some team members may have a preference for their *original* roles over performing business analysis activities (e.g., a developer is first and foremost *a developer*; and as such, he or she treats business analysis as a secondary task). The result may be that the business analysis activities get pushed aside or not performed altogether. However, there is a more significant issue about having business analysis skills *shared* by technical resources: when people *wear two hats* they may be more biased toward doing certain things over others, or taking certain approaches and performing work in a certain way. One of the most important characteristics of a BA is that he or she must be unbiased. The interest of the BA needs to be focused on value creation for the combined needs of the organization, key stakeholders, the project, the team, and the users. The BA must not take sides or have a preference as to what solution or approach needs to be taken. The BA can help address certain stakeholders' needs; but more importantly, the BA needs to be able to challenge those needs to ensure they are in line with the overall need to create value.

When the role of the BA is shared, it may be diluted with certain biases and the impartial, unbiased, and objective view of the BA may be *tainted* by technical or other considerations.

Scrum versus Agile

Fact: agile is an approach, an umbrella, even a state of mind—it is *not* a methodology. Scrum is a methodology—one of many methodologies that are recognized and positioned under the agile umbrella. This means that there are many ways to manage agile projects (depending on the methodology we choose). Some may not agree with the following statement, but it also means that there are multiple ways to interpret what agile means for each organization, situation, and context. No two situations require the exact same way of handling them; and no two projects will require exactly the same approach to agile. This means that we need to be pragmatic about being pragmatic (and agile is about being pragmatic). In North America, Scrum is by far the most common methodology under agile. There is one clear thing about Scrum: you are either doing it the way it says it should take place, or you are not Scrum. Scrum is black and white and leads to confusion when stakeholders, senior managers, and even Scrum practitioners fail to realize that there are certain situations where things have to be done differently.

An agile adaptation project we did at a large U.S. financial institution had multiple examples where stakeholders did not quite understand what agile was about and, at times, it appeared that some stakeholders lost sight of whether their goal was to be agile or to be successful in their projects. The reason was that some practices did not fit into the context of the organization, and when we proposed to *bypass* them, it triggered resistance that was rooted in misconceptions as to what agile was about.

A similar situation took place at a Canadian division of a large insurance company. The project sponsor, who acted as the product owner, was not going to be able to be available at frequent enough intervals. In fact, although the team *decided* on an iteration length of two weeks, the sponsor was going to be able to review the work only every three months. The solution was to bypass the organizational constraints, but team members and other stakeholders were lost on the difference between the methodology and the approach, and become too rigid in their attempt to be pragmatic.

In short, our goal is to be successful and, for the most part, agile methodologies offer answers to many of the problems that projects face. But at times, we need to remain pragmatic about it and remember that agile is a means to achieving success and not a goal in and of itself. Sometimes we need to bypass some of the rules it introduces—simply because they may not be suitable for certain environments and situations. And finally, we should also remember that being agile does not require anyone to follow any one methodology, including Scrum. Scrum is a way of doing things that is very specific, but it is not the only way to manage agile projects.

RECAP

This chapter deals with challenges to agile projects, environments, and organizations. It also offers solutions to overcome many of these challenges. Many of the solutions involve defining and appointing an agile BA—or at least identifying who needs to demonstrate business analysis skills when and under what settings. Many of the challenges stem from organizational processes, practices, and misconceptions; while other challenges come from within the team and the agile approaches. Some challenges are related to whether we should go agile for any given project, and in turn, when we consider agile, whether we are ready for it as a team and an organization and what needs to change so we can handle it.

Business analysis skills or even having a BA on an agile project is not the answer for all problems and challenges and will not serve as a *silver bullet* for all types of problems. However, there is a growing realization that business analysis skills are required and necessary to support agile projects—and recently, there is a growing realization of the need for having a dedicated BA within the agile team.

ENDNOTES

1. Stealth Methodology Adoption, Kevin Aguanno, *Multi Media Publications Inc.*, http://www.agilepm.com/stealth-methodology-adoption-book.
2. Agile Readiness Assessment is a proprietary approach developed by Genxus Consulting.
3. *A Guide to the Project Management Body of Knowledge (PMBOK® Guide)—Fifth Edition*, section 6.2.2.2 p. 152.
4. http://www.goldratt.co.uk/resources/critical_chain/.

3

ROLES AND RESPONSIBILITIES

All agile methodologies have specific views about the roles within the agile team and most are quite consistent with the roles of a Scrum team, on which we will focus our discussion here. The Scrum team has three roles: the product owner, the Scrum Master, and the team. We will go beyond providing a simple overview of the agile team's roles, and instead, we will explore the impact that the agile business analyst (BA) will have on the role and the way the agile BA can complement, help, and support these roles, while enhancing their performance for the benefit of the project and the team.

In this chapter we explore the role of the BA in agile environments, the impact that a BA would have on the other roles in agile projects, the changes project management offices will need to go through to adapt to agile environments, and the impact an attempt to adapt agile would have on different stakeholders.

AGILE PHILOSOPHY

There is a need to identify where business analysis skills are required in agile projects, by whom, and the timing of these needs. When necessary, there is also a need to find where a resource with a BA title best fits in. We should therefore recall some key aspects of the philosophy behind agile. Although the language in this chapter (and throughout the rest of the book in general) often implies the need for a person who has a BA title and who fulfills the role of the BA, the intent is to illustrate the need for these skills within the agile team; whether these skills are shared among team members or done by someone with a BA title whose specific role is to perform business analysis tasks.

Agile methods are based on just-in-time development and lean requirements analysis. They are business value driven and product oriented. Along with the myriad of benefits they offer, agile approaches may pose challenges if not followed properly and if not performed effectively. These challenges (discussed in

Chapter 2) can, for the most part, be overcome with the application of business analysis skills.

TRADITIONAL BUSINESS ANALYSIS ACTIVITIES

Typically, agile methodologies reduce the need for what we can call *traditional* business analysis skills and activities. The *traditional* BA role includes the following:

- *Exclusively leading the communication*: that is, the communication that takes place between the business and development teams.
- *Acting as the gatekeeper for requirements*: the requirements process is more intense since there is an attempt to finalize and baseline the requirements early on—pretty much before any work begins. This is somewhat unfair for the BA since he or she does not have the authority to make requirements-like decisions or to prevent business or technical changes.
- *Being a translator*: the BA must document and translate the business needs to the developers and the acceptance criteria to the testers. This *role* comes with the assumption that technical team members lack the right social skills to talk directly with the business—and the business is primarily viewed as unable to express their needs to the technical team members. The reason for these assumptions is not due to the belief that certain people are socially awkward, but it is more due to a track record where communication among technical team members and between technical teams and the business are often riddled with problems. These problems are not necessarily the fault of specific individuals or one group, but they are a reflection of a reality, where the business has a hard time expressing their actual needs and the technical people have difficulties articulating their understanding of what features need to be built and how.
- *Acting as a barrier or as a filter*: ensuring that technical team members do not share their full knowledge with the business. This commonly happens throughout the project when issues, problems, and defects surface. It is the BA who quite often ensures that no such *raw* information makes it back to the business.
- *Performing miscellaneous trivial roles*: due to the dynamic that develops, the structure of the requirements process, and the lack of clear understanding of roles and responsibilities, BAs often find themselves performing trivial roles. While it is not beneath anyone to do any type of work that helps the team produce value, these chores and activities are

well below the BA's skill level. These trivial activities are often time-consuming and they do not serve the intent behind having a BA in a project altogether—common examples include writing and documenting requirements, documentation police (ensuring team members document their work properly), version control police (ensuring that team members manage public documents properly), scheduling meetings, taking meeting minutes, and performing administrative chores.

These *traditional* roles are all driven by good intentions—such as reducing distractions, improving communication, preventing stakeholders from panicking for no reason, and protecting the team from misunderstandings. However, despite the good intentions, the result is not optimal:

- Team members often depend on the BA to finish their thoughts or complete their actions, instead of communicating directly and clearly when required.
- The BA is busy doing parts of the team members' job and *buffering* between different groups (business versus technical team and programmers versus testers). This is not a good utilization of the BA's time when the project could be benefiting from other business analysis activities that are not getting performed due to the lack of capacity of the BA. The waste also comes as a result of a duplication of effort between the BA and some team members, as well as potential gaps in areas that are not properly sorted out.
- The BA does not add sufficient or expected value for the project success, as he or she is consumed by mundane, non-value-added activities.
- Misconceptions develop around the true role of the BA, making the problem worse, because other people in the organization begin to utilize BAs for trivial activities, such as a personal assistant.
- With everyone used to the BA *cleaning up* and picking up the pieces, a culture of sloppiness is developed; team members neglect the quality of work that should be performed every step of the way.

AGILE BUSINESS ANALYSIS

The BA in an agile environment can help support the product owner. This refers to the ability of the BA to consolidate information, refer to subject matter experts (SMEs), seek clarifications, analyze and extrapolate information from data and reference materials (to make it easier for the product owner to make an informed decision), save the product owner's time by providing clear and concise information, and provide clarifications when the product owner is not available on a timely basis. These activities do not imply that the BA should replace the product owner or that the BA can make decisions on behalf of the product owner. The intent here is just to properly represent the mandate given by the product owner and help him or her support the project needs more efficiently. Here is a list of a few more benefits the BA can produce for the agile team:

- *Acting as a facilitator*: serving a critical need in the high degree of collaboration and interaction the team is dealing with
- *Support the team in consolidating feedback*: clarifying needs and requirements made by the business, help find the right person and type of help when team members hit hurdles along the way, help with figuring out the problem team members face
- *Support the development team*: close gaps in the elicitation, definition, analysis, documentation, and communication of requirements; help find the appropriate SMEs; help articulate business needs; consolidate and clarify technical feedback to stakeholders; help the estimating process
- *Support testing*: help in the process of articulating acceptance criteria, participate in writing test cases, consolidate information and facilitate the process of writing the test case, help figure out testing effort and estimates, help divide testing work and at times participate in testing
- *Help the Scrum Master*: as the BA tends to bring more product familiarity and potentially more technical orientation to the table than the Scrum Master, there is a need to coordinate and check if, where, and to what extent the Scrum Master needs the skills that a BA brings to the table
- *Serve as the Historian*: documentation skills come in handy in capturing significant information that may get lost in a team of purely technical people
- Make sure the latest information makes it out consistently and in a timely manner to all relevant parties

AGILE CHALLENGES WHERE A BA CAN HELP

In addition to the value the BA can produce in agile projects, there are many challenges that agile teams face that the BA can help address.

Getting the Development Team to Take Control of the Process

Volunteering is a key element of agile because agile team members pick their own stories based on a combination of business value, priority, project needs, and both team and individual skills. While agile environments are about encouraging ownership of the work, there are times where team members struggle to properly distribute the work on their own and some team members do not always pick the stories and requirements that they are most suitable to work on. While it is typically the role of the Scrum Master to oversee work distribution, a BA can help with this by providing more insights as to the priority, skills available, or team members' utilization.

Waterfall Thinking

As many organizations adapt different *flavors* of agile methods (we will leave the discussion on whether this is even possible for a different time and place), it is common to see agile projects that have a project manager (PM) instead of a Scrum Master, or even a PM alongside a Scrum Master. Beyond the question of whether it is beneficial for the organization, it may pose difficulties in the application of agile methods as well as open the door to duplication of effort and overlapping areas of responsibility.

A second group of issues may be introduced when there is no product owner, but instead there is a project sponsor. While the role of the product owner is often compared to that of the sponsor, they are not exactly the same.

A third type of problem that commonly occurs is the attempt to deliver agile projects with shared resources, as part of a weak matrix organization. When resources are not dedicated, it increases the chance that they may not report to the project on time or that they will be pulled out of the project for more urgent matters. Either way, it poses a threat to the team's ability to deliver value. These three examples all have a common denominator: they are triggered by senior management's *waterfall thinking*. This type of thinking is common since many senior members of management may not be familiar with agile methodologies and their thinking may still be driven by waterfall types of practices.

Having a BA (or someone who acts as one) in the agile projects can help reduce the impact of this behavior and help change it gradually toward a more agile approach. As senior management is often risk-averse—and as such, weary about agile practices—the BA can provide a cost-benefit analysis about agile practices and processes.

Despite the call to shift toward a more agile-oriented thinking process, in most organizations not all projects can or will be performed the agile way. Most organizations struggle in this matter given that both management and teams may confuse the two types of thinking (i.e., agile and waterfall), including processes and practices. The BA can assist in ensuring that agile thinking prevails, along with maintaining two sets of processes—one for agile projects and the other for waterfall.

Micromanagement

There are many forms of micromanagement. While some claim that agile is about micromanaging the team *without them realizing it*, those who say that agile turns into micromanaging are more concerned about problems related to middle management's resistance to change. This item is related to the previous point made here that many organizations are stuck in a waterfall state of mind, but here the reference is to members of the management team. Despite the intent to adapt agile methods and make organizational changes (including job titles and the introduction of processes and agile ceremonies), it is not easy to simply become a learning organization that embraces experimentation, collaboration, and (sometimes) failure. There are also challenges involved with empowering the team and turning it into a self-organizing and a cohesive team.

Organizational change is difficult on everyone, including senior management, middle management, and even the team management. The uncertainties often lead to conflict (within the team and with members of management), especially when there is conflict with management members' priorities and personal agendas. Another area that is prone to conflicts is when team managers (could be PMs or Scrum Masters) fail to allow the team to be self-directed and instead *assign* tasks to team members, thereby forcing completion deadlines that are often meaningless or unrealistic.

There is not much from a seniority perspective that a BA can do, but similar to the previous item, the BA can help walk members of management through the process, ensure the processes are clear, provide context and cost-risk-benefit analysis of the options and the advantages in following the agile processes, and help facilitate effective and open communication to reduce conflict and increase everyone's confidence in the process.

THE BA ROLE IN AGILE PROJECTS

Agile projects expect the BA to produce requirements faster and in multiple cycles throughout the project. In waterfall, the requirements process is primarily based on finding out what people believe to be facts before the development process begins, and then producing a series of long requirements documents. On the contrary, the agile requirements process relies on observation and trial-and-error—and the process leads to a much more efficient form of requirements documentation.

More specifically, BAs in agile projects should continue to champion and facilitate the requirements process, along with reinforcing collaboration with users. The key here is to work with those involved and get them acquainted and comfortable with the notion that ambiguity should be allowed until more information about the requirements is needed. It is hard for people to accept ambiguity; it is human nature to seek more information when things are unclear and to help veer toward a sense of comfort. In agile projects, there is no need to seek more information about requirements at the early stages of the project since many of these requirements may change or get canceled later. The BA is the ideal role to facilitate the agile requirements process and provide the appropriate hand-holding while dealing with ambiguity. With the help of the BA, the focus of the stakeholders and team members should shift from attempting to look for details to identifying missing requirements.

Regardless of the agile methodology selected, there is an increased focus on the early definition of acceptance criteria. Some methodologies (i.e., test-driven development) put even more emphasis on the definition of acceptance criteria since it, in fact, couples with the creation of test cases before the development effort takes place. This style of programming ties testing, coding, and design together and focuses on first defining acceptance criteria and writing test cases for these criteria—and only then build features that correspond with the test cases that are already in place.

Being Ubiquitous: Creating Shared Understanding

Declaring that an organization is adapting agile methods and following agile processes and ceremonies in projects does not automatically mean that all team members, stakeholders, and users become collaborative, work effectively together, or do what they are supposed to. Misunderstandings still take place, intentions and needs are often miscommunicated or poorly articulated, and technical team members may not always build the right thing or build it the right way. For these problems and issues, the agile BA often needs to wear different hats by putting him/herself in the shoes of different roles to better understand other points of view. These shoes include those of the designers, developers, testers, and as mentioned earlier in this chapter, even the product owner. Helping team members and stakeholders understand each other also involves the need to mentor technical team members on how to effectively work with business stakeholders and, at times, assume the role of the product owner for clarifications and in order to expedite things with a sense of urgency.

In addition, the BA may also need to stand in for users when there are issues with the timeliness and the level of their involvement and availability. The term *stand in* refers here to the BA's need to do it only on an ad hoc basis, when required—and not on an ongoing basis as happens in waterfall projects.

Overall, the agile BA needs to be ubiquitous, in the sense that he or she gets involved wherever and whenever business analysis skills are required—all in the name of creating a shared understanding of how the product needs to look and what it is supposed to do. The involvement can be in the form of facilitation, mentoring, hand-holding, role representation, or playing devil's advocate—with virtually any role in the agile environment.

WHAT TO EXPECT OF THE AGILE BA

Beyond the activities that the BA needs to perform in agile environments, there are some characteristics that the agile BA needs to have. It is important to note here that these characteristics refer, for the most part, to a resource whose title is a BA in agile projects. It is possible to find these characteristics among the technical team members who act as part-time BAs, but it will be hard to find these characteristics consistently among team members, or having all of these characteristics associated with any one individual team member.

Before going over specific and relevant characteristics for the agile BA, it is important to recall Stephen Covey's Seven Habits of Highly Effective People:[1]

1. Proactiveness,
2. Goal-oriented (think with the end in mind),
3. Prioritize (put first things first),
4. Win-win thinking,
5. Clarity: seek first to understand, then to be understood,
6. Synergize, and
7. Add value (sharpen the law).

As these habits are applicable to multiple situations, anyone in an agile environment—especially the BA—needs to focus on these habits. The following list of attributes and characteristics for the agile BA is comprehensive and it might be difficult to find one individual who demonstrates all these traits.

Such a variety of skills is easier to find among different team members who can complement each other; and if there is sufficient capacity within the team, each team member can act partially as a BA in an agile project. There is, however, a need to look for these characteristics when considering an individual to act as a BA in an agile project. These are also areas for potential professional development for people who embark on a career as a BA in agile project environments. In addition, many of these characteristics are in line with those expected from a *regular* BA who works on waterfall projects. The list in this section, however, has some additional nuances for the agile BA.

Focus

The agile BA's main focus is to help the organization solve a problem and ensure that all activities are value-added ones. It is normal to expect that team members get tangled in the technical aspects of the solution, but the BA must always keep the bigger picture in sight, which is to resolve the business problem at hand. With an eye on the big picture, the BA can help ensure focus on solving the actual problem as well as ensure that the project does not attempt to address other problems. When projects cease to produce value for the intent for which they were chartered, it ends up causing waste and increasing cost for no valid reason. Focusing on the goal (keeping the end in mind) is part of the agile BA's ability to see the big picture and to apply systems thinking while facilitating attention to detail where necessary.

Innovation

There are different contexts for the word *innovation*; and when we say that the BA must have innovation skills, it is not the same as when we say it about the developers. Technical team members need to be innovative as to how to build a solution that addresses the true needs of the customer, which is technical innovation. The BA needs to be innovative with several areas that are mostly process driven:

- The approach the BA takes to engage stakeholders and elicit information and requirements
- Process improvement ideas
- Communication methods
- Incorporating early and sufficient involvement of the testing team
- Contribute to the introduction and utilization of automated testing and deployment tools
- Facilitate clear project reporting
- Cover for any stakeholder and team role that are not readily available for input or feedback
- Proper and timely capture of stakeholders' feedback and performance of backlog maintenance activities
- The BA also needs to be innovative in helping turn the team toward becoming self sufficient and (in some form of irony) help reduce the team's dependency on the BA
- As we expect the development team to be innovative in building the solution, the BA needs to be one step ahead, pursuing new ideas and approaches to address the business problem and improving processes, and to introduce innovative measures to diagnose business processes that need improvement

Proactiveness and Initiative

- The BA has to be on guard to ensure that all parties remain focused on actual business needs so that the solution does not veer away from the requirements it needs to address

- The BA has to review items in the *shadow backlog* (i.e., unprioritized and unintimated stories driven by lower priority changes and needs, as well as deferred defects) and bring up stories that are suitable (based on their importance, as well as progress made by the team and the upcoming iteration's goals) to be incorporated in the upcoming iteration. These stories need to be corroborated, prioritized, and estimated by the team and approved by the product owner

An additional element of proactiveness is the ability to anticipate. Agile approaches call for planning only at a high level with pulses of detailed planning one iteration at a time (i.e., two to four weeks), and therefore, it cannot be expected that we will know any details (including specific features and functions) for the period of time that is beyond the next iteration. With that said, the BA is familiar with the business domain and should be able to anticipate impact that may occur across the organization as a result of actions, activities, and even features that will take place beyond the immediate planning horizon on the upcoming iteration. It is not expected that the BA predicts the future, or that the BA will know play-by-play what is expected to take place in the future, but the BA must (and is expected to) have the ability to anticipate events, reactions, and impact areas further into the future. The need to anticipate is to ensure that the product that the team is building produces value for the entire organization and not only for one domain.

The need to anticipate may apply an opposing force on the BA: while the BA works closely with the team, being proactive and maintaining the ability to anticipate may slightly detach the BA from the team, so he or she can maintain a sufficient level of independent thinking in order to evaluate the true impact the product has on the organization—while still maintaining objectivity. If this is done properly, it can provide the organization an insurance policy that keeps an eye on long-term benefits.

Context

Context is a broad and loose term to describe a characteristic of the agile BA. It refers to sufficient understanding of anything that is around and that can impact the business needs, the project, the processes, the product, and the creation of value. The agile BA needs to have sufficient understanding of the product in order to be proactive and suggest which items should be considered to be included in the upcoming iteration.

Understanding of the organization is required for reporting lines, resource allocation, measuring resource skill level to support resource story selection, and knowing team members' areas of expertise. It is also important to know escalation lines and procedures to help bring the right type of support for technical issues and impediments.

The BA also needs to have a clear understanding of the business needs to provide meaningful support for the product owner, as well as to help facilitate the stories and requirements prioritization process.

The BA has to be familiar with processes that impact (or support) value creation and the work of the team. Although the BA may not be the owner of these processes, it is important to know who owns them, the rationale behind their structure, and which parts can be challenged by the team in pursuit of efficiencies.

Additional items include information on a less formal level—as it can help if the BA is familiar with the team members and stakeholders, including their personalities. Industry and regulatory knowledge can also be beneficial.

A quick review of the breakdown of the word *context* here shows that it will be highly beneficial to have a BA brought into an agile project from within the organization. Bringing a new BA into the project from outside of the organization (even if experienced) may not allow him or her to maximize the value that the BA can add to the project and the organization.

Business orientation also falls under the broader scope of the word context. The need for value creation (a production-quality or releasable product increment in every iteration) means that every feature and functionality that the team builds must address a business need. While everyone on the team needs to be business oriented, this alone may pose a challenge in some environments since when everyone is in charge of something, it may not be owned by anyone in particular. In such cases, it is often easy to veer a little sideways and start producing products and features that do not properly address business value—because there are so many moving parts and distractions. With heavy reliance on the product owner for business justification and direction, the agile BA must be business oriented and constantly stand up for activities and features that are driven by adding business value. The BA must always ensure that solving the business problem or addressing the opportunity is at the top of the team's priority list and that the solution that the team needs to build is the means to add value and not a goal in and of itself.

Critical Thinking

Critical thinking is something that all team members should possess; but when people get busy and distracted, there is a need to have someone around who is skilled at critical thinking. The entire role of the agile BA revolves around critical thinking since the BA needs to have an eye for waste, opportunities, savings, and efficiencies. Critical thinking can generate challenges, change, and even a healthy dose of conflict to ensure that people do not settle too deeply into their comfort zone.

While it is important for the BA (and for all others) to demonstrate critical thinking, we need to keep in mind that too much of it may be perceived as non-confirming, and some even view people who demonstrate critical thinking as troublemakers. In fact, in this day and age, many organizations mistake critical thinking to be a threat since it may be interpreted as people's attempts to stir trouble, promote agendas, or challenge structures, processes, and stability. The BA must strike a balance and not slide too far on the path to critical thinking. In addition, the innovation, proactiveness, and critical thinking must remain confined to the context of agile and the team.

A Balanced Approach

The BA must navigate his or her conduct along a fine line between challenging processes and demonstrating critical thinking on one side and maintaining a level of stability and comfort zone on the other. Since agile is about reducing risks, the team performs within a stable cyclical set of processes in order to produce a production-quality product increment in every iteration. The fixed time-box approach in agile projects is intended to create a comfortable rhythm for the team to build and produce the product, therefore the BA needs help from the Scrum Master or the PM to guard this comfort zone and help enable the team's performance.

Storytelling

The fact that it is an agile project does not mean that all information is clear and streamlined. There will still be difficulties in identifying actual needs, articulating them, and conveying them clearly and in the right context. In addition, through different stages of the project, it is likely that key stakeholders may not be readily available or they may lack the clarity to express their needs. On the team side, there may also be challenges in reflecting on these needs and converting them to the envisioned features and functionality. It is therefore important to have someone (i.e., the BA) with storytelling skills who can provide another set of ears to listen to client needs and help the team convert them into requirements and user stories that will meet those needs. The storytelling skills can also be useful when validating requirements, communicating with stakeholders, and conveying messages to and from the product owner.

Leadership

Leaders are not appointed to their job and one does not need to be senior in an organization to demonstrate leadership skills, therefore it is reasonable to expect that the agile BA is a leader. As such, the BA needs to facilitate effective teamwork and ensure that the team focuses on teamwork rather than individual performance. The BA also needs to initiate changes (focusing on processes) and improvement ideas across the organization. Combining leadership skills with innovative measures and effective communication will help the BA build credibility with the team and business stakeholders. This leadership factor is not achieved through authority, but through influence, facilitation, and communication skills.

On a side note to leadership skills, the BA has to be confident and comfortable when talking to anyone involved with the agile project, including senior executives (about business matters, the marketplace, the product, and the competition), external stakeholders (about needs, updates, changes, and reporting) and the team (about requirements, user stories, and changes).

Adaptability

Adaptability is something that everyone involved in an agile project needs to have. When it comes to the agile BA, adaptability is not only about the evolution of the work and the product, but also about changes to the BA role itself that may take place throughout the project. With a wide variety of changing needs and potential focus areas, the needs for business analysis skills may change as the project progresses—requiring the BA to be flexible and to demonstrate different sets of skills at different times. Needs may include any of the following:

- Help with and facilitate the requirements process
- Help elicit and gather information
- Support the development team
- Assist with the testing process
- Improve business processes
- Facilitate reporting, modeling, and documentation
- Assist in decision making
- Help design user interfaces
- Facilitate communication to and from the product owner—at times acting on behalf of the product owner if and when required
- Support the Scrum Master or PM with stakeholder engagement and organizational change considerations

The agile BA must always keep in mind the big picture since the BA is not confined to a single or isolated area. Therefore, the BA needs to ensure that actions or changes that take place in one area do not have an adverse effect elsewhere; and if there is an impact, it is properly addressed or communicated even across framework and organizational boundaries.

Communication Skills

Last but not least, the BA must demonstrate all-around, excellent communication skills. Many of these skills have been implied or included with the skills mentioned in this section, but other aspects of excellent communication skills include articulating a message, documenting and handling matters, escalating issues efficiently and effectively, actively listening, and being flexible. This enables the BA to change his or her style appropriately when dealing with different stakeholders and different communication needs.

An important component in developing the flexibility to handle different stakeholders is through empathy. The BA needs to exhibit empathy and understanding and act as an intermediary who is, in part, mediator, moderator, and an island of composure in the midst of the storm of different views, opinions, and approaches. While doing all that, the BA must keep an eye on efficiencies, business value, and the overall impact on the organization. Empathy can go a long way in helping stakeholders and team members to open up, creating for them a safe zone to communicate and helping establish trust all around. Empathy can also help build relationships within and outside the organization.

AGILE BA RESPONSIBILITIES AND ACTIVITIES

The application of business analysis skills in agile projects can be broad and span across many aspects of the process. To maximize the benefits the BA produces for the team, the project, and the organization as a whole, there is a need to identify areas of needs and define the focus of the BA's work. As the project progresses, focus areas will need to change as required, but the initial set up can help ensure that the BA adds value where it is needed the most, and does not meddle in areas that do not require specific business analysis skills. The set up needs to be done by the Scrum Master, or the PM and BA, and if required with help from the product owner.

User Stories

Activities that the BA should be involved in include anything related to user stories—facilitating the process of identifying user stories and ensuring information gathering and elicitation techniques are suitable and utilized properly for the project's needs. The BA may lead some of these activities or help set them up for team members. The BA can further assist by helping describe and document user stories—specifying minimum viable functionality and acceptance criteria (i.e., the *definition of done*) and adding them to the backlog based on the product owner's mandate.

If required, the BA can also assist by working with the stakeholders to help prioritize the product backlog of user stories and ensure that all considerations have been taken into account for the prioritization process. In turn, when stories are selected to be built in an iteration, the BA will be there to help the development team understand the details of the story and seek more information if required (by identifying and approaching the relevant stakeholders and SMEs). To further help the team understand the detailed requirements, defined tasks, and estimated effort, the BA can help communicate the details to the development team and, where required, provide additional supporting information based on the team members' needs. The supporting information may be provided in any format (formal or informal) and may include any combination of narratives, models, diagrams, or storyboards. The BA also helps facilitate the documentation and retrieval of information through storage, access, and version control of documents to ensure accuracy and process efficiency.

The BA is not supposed to write individual stories, but rather ensure that the stories that are written are clear, concise, easy to understand, and easy to implement by the development team. Even without direct involvement in the actual writing of stories, the BA can help provide and facilitate the criteria for what constitutes a good story so that whoever writes the story has the ability to determine whether the story is clear enough or whether there is a need to seek more information. Writing good and effective user stories is not a skill that should be taken for granted—and the BA can help ensure that each story not only has a clear functionality, but also that the *why*, or the *so that* section of the story, expresses the business value that the story is intended to provide. Sometimes the BA's point of view will provide more context than that of the developer.

Be There for the Team

The BA not only helps the developers but also provides valuable support to the testing process: this includes articulating acceptance criteria and the definition of *done* (ensuring that the development team understands what the minimum viable solution should look like) and communicating the acceptance criteria to the testing team. The BA may also assist with the creation of test cases based on the acceptance criteria, and at times, the BA may also take part in executing the test cases and performing other testing activities.

The BA should also assist with ensuring non-functional requirements are taken into consideration, as well as making sure that the creation of the documentation that is necessary for regulatory purposes and for future maintenance is taking place.

Facilitation Skills

There are many workshops in agile projects and they lift the set of facilitation skills to become one of the most important roles of the BA. In some environments, BAs with strong facilitation skills can become Scrum Masters, while others continue in the capacity of a BA—to take command of meetings and lead requirements and other types of workshops and sessions that require facilitation. Strong facilitation skills can also help overcome situations where the BA does not have strong technical skills since the facilitation skills can help engage team members, stakeholders, and SMEs in search of answers to technical issues and questions.

Help with Agile Ceremonies

When it comes to agile ceremonies, the BA can help team members ensure that reporting is always up to speed, that accurate updates are sent to the Scrum Master at the end of the day and prior to the Scrum stand-up meeting, and that escalations and troubleshooting is taking place as required. The BA can play an instrumental role in the team's preparation for the iteration review and the demo. Beyond helping consolidate information and clarifying needs for the demo, the BA can, at times, role play on behalf of the product owner as part of the iteration review and demo prep process. During the actual demo, the BA will help by documenting feedback and changes to the product backlog and providing clarification that all participants are on the same page.

Last but not least, the BA can help the team organize thoughts and ideas for the retrospective meeting and then during the meeting, take notes, ensure that the lessons are clearly stated, help identify specific actions, and ensure that decisions are clear. During the upcoming iteration, the BA will facilitate the process of applying the lessons and capturing any type of feedback around the benefits, impact, and side effects gained from the lessons. Before the next retrospective, the BA will engage team members and consolidate information and provide a report during the meeting as to the items previously identified and applied as well as to the level of success achieved. Beyond the team's application of the lessons from the previous retrospective, the BA should keep an eye out for any other challenges, waste areas, issues, unrealized efficiencies, and process improvement opportunities.

A review of common BA tasks shows that, fundamentally, the tasks of the agile BA do not change much from those of the *traditional* BA role, but the format, the timing, and the environment definitely change. Another thing that changes is the impact that agile has on other roles, including the ambiguity and the cyclical pulses of requirements, planning, building, and testing activities.

Decisions/Decisions

The team should be able to determine a sufficient level of details and breakdown of user stories in order to determine when there is sufficient information and when more information is needed. This includes the ability to determine which stories will remain as epics. The BA can help in this area by facilitating the creation of criteria and the search for additional information when required. When stories and epics follow a logical functional hierarchy, it provides a mechanism to better understand the relationship of the stories and epics to each other and to satisfy the overall business goals by keeping in mind risks and business value.

Since mandates and themes provided by the product owner may, at times, be vague, the BA can help the team identify related user stories and epics, group them into themes as necessary, and if required, document the interrelationship and associated business process flows. When dealing with a large and complex project, there may also be a necessity to integrate the needs of related projects, as well as the needs of different stakeholders, to produce an overall solution.

Questions to Ask

In coordination with the Scrum Master or the PM, the BA should always be prepared to *think outside the box* or to play *devil's advocate*. The intent is not to always challenge everything, but to help team members and stakeholders avoid group think and reduce the chance of locking themselves into paradigms and misconceptions. The following list of questions is intended for BAs to consider and tap into if and when required—with some questions best utilized by being floated at the back of every BA's mind:

- Can you explain that to me?
- Is there another way to do this?
- How can we do that better?
- What comes next?
- What benefit will this produce? To whom?
- What are we missing? (This is not a functionality question, but a process-driven one that causes us to contemplate whether there might be another hidden consideration or another way of doing this.)
- What is your responsibility?
- Why is it important and what makes it important?
- What is going to be the end result and how is it going to look?
- What type of help do you need?
- What additional information do you require?

Additional Considerations

Unlike in *traditional* environments—where the BA leads most engagement, communication, and collaboration activities—the BA in agile projects needs to actively encourage and facilitate these aspects so that the team members perform them. As the BA continues to facilitate these processes, as opposed to leading them, it will provide additional capacity for the BA to improve the interaction with users, developers, testers, and other stakeholders—especially through the multiple cycles of the requirements elicitation and gathering processes.

In addition, since agile project environments are typically more inclusive, open, dynamic, and animated than those of a waterfall environment, it may add to the confusion and generate a sense of unease among some team members (especially introverts). Although the dynamic environment produces multiple benefits, when individual team members feel uncomfortable it may shut down their creativity, communication, or ability to collaborate. The BA can help identify these challenging situations, produce ideas on how to overcome them, and provide additional support to those who need it. The support can come in the form of finding quieter and less hectic areas for some individuals to retreat to in order to help them express themselves more effectively.

The BA can also add value by always being there for team members to ensure a quick turnaround or a resolution for small issues that may slow them down. Such issues commonly include prioritization, challenges in understanding requirements, issues around estimating, or technical problems. While in some cases the BA can provide answers, the help that the BA provides may also include proper escalation or direction as to who (i.e., which stakeholder, SME, or team member) to engage in the matter.

The fact that the role of the BA in agile environments is not fully defined can provide flexibility and the potential for growth as to where the BA can add value to the team. While small projects may do perfectly fine without the specific role of a BA or even a breakdown of business analysis skills among team members, large and complex projects will need a clear breakdown of business analysis skills as well as a definition of the role of the PM. Lack of clear definition of business analysis skills can inflict even more damage in cases where the product owner is not sufficiently available or when the product owner lacks the required business analysis skills. Even when the product owner is present, large projects can benefit from a BA to support the product owner.

ROLE IMPACT

Before discussing the impact that the BA has on the various roles in the agile project, let's review some key differences in the way the team handles itself in agile projects since the agile team is more self-directed and self-managed than teams in *traditional* projects.

Team Differences

Along with the shift in the way all agile team members perform their work, there is also a significant role impact on the BA because there is an expectation for more collaboration, initiative, and proactiveness in agile environments. Table 3.1 highlights the characteristics of the team and the BA.

Table 3.1   Differences between closely managed and self-directed teams

Closely Managed Teams	Self-Directed Teams
More reactive, taking directions	Proactive, taking initiative, members pick stories to work on
May focus on seeking individual rewards and at times even compete with each other	Contribution to the team and team success is the primary goal, to be achieved through collaboration and cooperation
Comply with processes, little propensity to challenge process	Continuously look for process improvement and efficiency ideas
React to issues, challenges, and emergencies	Take proactive steps to take preventive action
Focus on low-level objectives and tasks	Focus on creating value and on the solution
Rapid shifts from pre-deadline intensities, to long *lulls* in between milestones	Deliver a consistent and sustainable level of rigor over time with a manageable workload through closely spaced milestones (iterations)
Tend to wait with issues until close to deadlines or until symptoms are showing	Proactively and systematically address issues in a timely manner
Avoid changes and seek to maintain status quo	Seek improvements and apply lessons regularly
Try to build a sense of confidence in long-term estimates early in the project	Add value by providing high-level estimates early on, supported by multiple timely rounds of validation one iteration at a time
Rely on higher authority for resolutions, at times through unclear escalation procedures	Self-directed in addressing issues and consistently escalating when needed

Team Impact

Beyond the impact that the BA may have on a specific role, there is also a general impact that the BA may have. In larger and more complex projects, many stakeholders may not have a chance—or may not be available—to get involved or to have the full and proper ability to express and represent their needs. In situations like this, the BA can help ensure that the needs of the stakeholders get represented. The fact that the BA is reaching out to stakeholders also helps with the need for continuous contact with stakeholders. While in waterfall environments phases are very distinct and the majority of the contact with external stakeholders takes place during the requirements phase and then upon releasing the product, in agile projects there is a need for continuous stakeholder involvement and engagement because the project evolves in small bursts of requirements, estimates, and planning in each iteration.

Systems Analyst

In many waterfall environments, it is common to find more than one person in the role of BA with some BAs focusing on the business side and others on the technology side. In agile projects, if there is a dedicated BA, he or she is primarily focused on the business side and not on the technical domains. This reality does not necessarily create a gap on the technical side since many agile team members also act as BAs, but with a more technical inclination based on their role. If there is a gap in the attempts to coordinate between the business and technical sides, it leads to the need to bring back the systems analyst (SA) role. With the BA performing the non-technical activities, the SA brings back the old-fashioned role of understanding impacts across systems, dependencies across processes, and the requirements of the production-support teams.

The SA role does not create a duplication of effort and is required only when such analysis skills are not readily available among the technical team members. However, the SA role can also help when there are challenges that are triggered due to the lack of technical documentation since the agile team primarily relies on the needs and views of stakeholders that are expressed in meetings. With potential misunderstandings, the SA can help the team make the right technical decisions and recommendations—or look for additional technical help from SMEs. If performing the role effectively, the SA role can become a proactive beacon that facilitates the pursuit of technical information sufficiently in advance, address some of the challenges with the BA, and at times, even provide subject matter expertise. In many agile environments, the developers migrate into SA roles since they are frequently asked to perform spike (analysis) activities that go beyond the application area that the team is working on.

End-User and Non-functional Requirements

Some agile teams shift their focus from business requirements to end-user requirements. This often involves an increased focus on interviewing users, where the *traditional* skills of the BA come into enhanced play—with a focus on elicitation, facilitation, documentation, and communication skills. When teams shift toward a more end-user requirements approach, there is an enhanced need for the BA within the team.

Another reason to include a BA is driven by the tendency of some agile teams to *forget* about non-functional requirements. While it is not common, such problems can be expected since the structure of user stories naturally focuses on functional requirements. In such cases where non-functional requirements are left out, a BA can help in measuring, eliciting, or gathering information related to a series of characteristics and needs that are captured by non-functional requirements. Things like accessibility, maintainability, and other words that end with "ility" are important for product quality; and although it is not the ideal use of the BA's time, the BA can help ensure these things are looked after. Table 3.2 provides a list of common non-functional characteristics. The BA can help with these attributes by helping the team find the right expert if needed.

Table 3.2   Non-functional attributes—the "ilities"

Attribute	Explanation
Availability	Probability of performance; the product will perform as required, when required
Maintainability	Mean time to repair
Affordability	Return for quality; the ability to develop, acquire, operate, maintain, and dispose of a product
Producibility	Technology required
Flexibility/ Extensibility	Expected variable use; ability to use the product for different purposes, capacities, and conditions; is the solution designed to support new functionality in the future?
Operability	Expected conditional use; ability of product to be operated for specified time, under specific conditions, without significant degradation of the output
Reliability	Mean time between failures (MTBF); calculated by predicted and actual MTBF
Usability	Effort expended to use; the ability of a product to perform its intended function for the specified user under the prescribed conditions
Social Acceptability	Environment and safety; compatibility between product characteristics and the prevailing values and expectations in the jurisdiction
Capacity	How much traffic must it be able to handle?
Deployability	Will the solution need to be deployed all at once or in components and phases?

Privacy and Security	Is the solution designed to account for specific issues related to data security and integrity, customer access, and data awareness?
Scalability	To what extent must the solution be expandable to handle increased throughput?
Interoperability	Does the solution need to interface with other products and if yes— how?
Recoverability	How quickly must the product be up and operational after an outage?
Response ability/ Response time	How quickly must the product respond after a request is made?
Internationalization	Does the solution need to support multiple languages, time zones, or standards?
Portability	Does the product need to work on other platforms (e.g., hardware, operating systems)?

The Product Owner

There are many business analysis aspects within the product owner's role. In fact, most of what the product owner does, except the decisions that he or she makes for the business and on behalf of the users, is closely related to business analysis. When the product owner is not sufficiently available (i.e., time, involvement) to work with the team, the BA can step in and partially act on behalf of the product owner. When it involves handling clarifications and streamlining communication, this is a great way for the BA to add value for the team. However, if the BA has to start guessing and making unfounded assumptions or decisions on behalf of the product owner, this is when problems begin. The team and the stakeholders may get frustrated with the sudden increase in the BA's power, and the BA may lose site of the true meaning of the business analysis function in the project. Beyond these power and politics considerations, it is most likely that the BA does not have the business context, knowledge, understanding, and acumen to fully act on behalf of the product owner. In addition, it is obvious that the BA cannot think on behalf of the product owner.

Testers

The role of the testers in an agile project increases—at times, dramatically— compared with a waterfall project. Testing in an agile environment is so important and so different than it is in *traditional* projects that there is an entire chapter in this book (Chapter 8) that deals with it.

Testing and quality assurance (QA) are a set of activities that are intended to ensure that the product meets the customer requirements in a systematic and reliable fashion. In agile, it is expected that testing is the responsibility of all members of the technical team, and not only that of the testers. QA and testing involve all of the activities that serve to ensure quality and product correctness throughout the development of the product. With agile, testing is done early in every iteration and the testers take a more integrated and team-oriented approach than in waterfall environments.

A generic agile test strategy includes:

- The development of a test strategy along with the release plan,
- Planning and performing QA activities that take place in each iteration,
- System testing (known as a *hardening sprint*, to fix defects from previous iterations), and
- Release QA activities.

In many agile environments, the BA helps the testers at any stage and at various levels—from defining the acceptance criteria, to writing the test case, and even to performing testing. The help can involve actually doing the work or just facilitating processes, helping build test plans, or coordinating the effort. With the increasing workload of agile teams, many BAs find themselves getting progressively involved with the testing process to the extent that those teams practically merge the role of tester and BAs. Although it may pull the BA away from supporting other parts of the project, it can work well if the BA is a strong tester.

Technical Writers

In many environments—both agile and waterfall—BAs are often utilized for supporting activities that do not add value in and of themselves, but rather when the BA performs these activities, it simply saves time for other team members. Examples for these types of support activities include documentation, technical writing, and building training materials. While having the BA performing these types of activities can help free up time for other team members, this is an old-fashioned view of the role of the BA as a *glorified butler*, or a generalist, and reduces the amount of value that the BA can actually produce for the project.

Knowledge Management

According to the Project Management Institute, there is an increased awareness of the area of knowledge management. "The increasingly mobile and transitory workforce requires a more rigorous process of identifying knowledge throughout the project life cycle and transferring it to the target audience, so that the knowledge is not lost."[2] The term *knowledge management* refers to knowledge and information that are created or reused to foster collaboration in the work environment. It includes knowledge sharing and knowledge integration that is collected from different domains, not only knowledge that is needed by the technical team to write new code. Agile projects need to produce knowledge in the areas of user support tools, production support handover materials, and information for business users.

Whereas the task of creating, retaining, and reusing knowledge requires continuous focus throughout the life of the project, there should be resources who perform this work—but it can be supported and facilitated by the BA in coordination with the Scrum Master or the PM. Beyond the creation of the knowledge, the team needs to ensure that the knowledge format can outlive the project.

Scrum Master, the PM, and the Team

The Scrum Master (or the PM) is the main coordinator of the agile project, focusing on removing any impediments, blockers, and interferences that may prevent the team from making meaningful progress. The Scrum Master leads the agile project team and is in charge of leading meetings, facilitating conflict resolution, enforcing rules, and protecting the team from distractions.

There are clearly multiple touchpoints and potentially many overlapping areas between the Scrum Master (or the PM) and the BA. As a result, it is critical that the PM and the BA establish rules of engagement, role definitions, and boundaries—so they can both focus on their strong areas and maximize the production of value for the team through collaboration, coordination, division of work, and open communication. As conditions throughout the project can change quickly, so must the need to coordinate both roles' efforts and to change the division of work and areas of responsibilities. Although the BA has an important and significant role in the agile project, it is all done under the umbrella of responsibility for the project—provided by the PM. Both the PM and the BA provide support, help, facilitation, and removal of impediments for the team; and the need to coordinate these efforts is critical so there are no gaps or duplication of effort. With that said, the division of work between the PM and the BA is fairly easy to make; the PM leads project aspects, change management, and most of the stakeholder expectations management, while the BA is more involved on the product side, requirements, modeling and creating business value, and supporting the technical work of the team.

No BA? Team Members Are Part-BAs

When there is no dedicated BA in the agile project, the PM facilitates the effort to help the team identify who will perform which business analysis activities and areas. When the technical team members are part-BAs, there is a need to ensure that all business analysis-related activities are coordinated with the PM and that there is a clear definition of the expectations from each team member.

There are a few challenges that may surface when there is no dedicated BA within the team:

- Multiple team members perform business analysis activities, which may lead to misunderstandings, duplication of effort, and gaps.
- Technical team members tend to be first and foremost technical and only then, BAs. As such, the business analysis activities and areas of focus may be pushed to the side because the technical role will take priority.
- With several people performing business analysis work, there is a need to coordinate who does what and ensure that team members have the right skill set for the business analysis tasks.
- There may be lack of ownership of business analysis work since no one individual is in charge of consolidating it.
- The Scrum Master or PM needs to establish multiple formal touchpoints to coordinate the team's effort in regard to business analysis activities.

Despite the potential challenges when there is no dedicated BA in agile projects, if the business analysis work and areas of responsibility are addressed and properly distributed, it can serve as a strong advantage for the team. It will improve the team's performance, the project will benefit from the equivalent of an additional set of eyes in any given situation, and each individual also acts as a BA.

PM and BA Role Collaboration

The collaboration between the Scrum Master (or PM) and the BA needs to take place from the very start of the project and continue throughout. To be clear, the PM serves as an SME in the agile, change, and project management processes and the BA (whether it is an individual with a BA title or the skill is spread among the team members) is an SME in business analysis and requirements management processes. Together the PM and the BA facilitate the production of realistic integrated plans that cover all stakeholder needs, help manage expectations, support the prioritization effort, and oversee the building of the product increments and the production of value. Many of the activities that the PM and the BA perform in *traditional* projects will now be performed by the team in agile environments—including most reporting, communication, dependencies, and impact areas.

Another area that makes the relationship between the PM and the BA in agile projects easier to manage is the reduced amount of inherent conflict between the two roles. In a waterfall project, there is constant friction between the PM and the BA because the PM typically focuses on schedule and cost performance, while the BA is primarily concerned with scope (requirements) and quality. This conflict is not an issue in an agile project since there is no uncertainty about trade-offs among the competing demands—all trade-offs are addressed through prioritization of scope items, and any time something happens that reduces the team's velocity or capacity, a lower priority scope item will be removed from the iteration and bumped into future consideration. Time, cost, and quality are always delivered, which makes it easier for all involved to produce value for the business.

BUSINESS ANALYSIS AND PROJECT ACTIVITIES

The BA oversees, leads, or facilitates many project activities, and through coordination with the PM or Scrum Master, the BA can produce a long list of benefits for the team and the project. This section provides a review of some of the project activities that the BA impacts.

The Agile BA and Facilitated Workshops

Depending on the workshop, the leader may be either the PM or the BA. When someone leads or facilitates a session, it does not mean that he or she does everything in the session. When dealing with a facilitated workshop, there are a few roles that add value to the session; and the meeting leader (e.g., the BA) needs to get team members to fulfill these roles. Common roles include the following:

- *Owner*: sets the objectives and deliverables of the session. This could be done by any of the project or solution development team roles—or the product owner.
- *Facilitator*: manages the process of the session, establishes rules, and ensures the flow of information. This role has to be done by a neutral and impartial leader and enabler who has no direct stake in the content of the meeting or its outcome—besides ensuring that the outcome is achieved. It is common to see the BA performing this role. The meeting facilitator does not need to be a senior individual with formal authority, but rather a leader who is accepted by the meeting participants. There are often challenges with the BA's ability to contain the session's dynamics, especially when senior stakeholders participate in the session in an attempt to *override* the facilitator's authority.
- *Scribe*: the scribe supports the facilitator and records results and actions, decisions and/or issues, as well as action items. The scribe needs to have the appropriate knowledge and context, and he or she needs to take proper and clear notes. However, since the scribe will be occupied with the notes, this person should not be a technical person who has a stake or an active role in the meeting's content and outcome. This stipulation makes it more difficult to find a scribe. This role can be performed by the facilitator if he or she has sufficient capacity to do it. There may be a

need for more than one scribe in a workshop depending on the session type and the amount of information that flows.

- *Participants*: team members and stakeholders who participate in a workshop must add value to the purpose and flow of the meeting; have the knowledge, skills, and experience that are relevant to the session; and be empowered to take part, make a contribution, provide information, and make decisions. The BA and the PM or Scrum Master are not active participants in such sessions.
- *Observer*: this is an optional role that could be done by a member of management, an auditor, or a trainee. This is a non-contributing role and he or she must not distract the participants. It is a good practice to have them explain their presence to rest of the participants.

Prior to the session, the facilitator circulates information to participants—enough in advance so that the participants can review the material. The facilitator also sends out an agenda detailing all of the logistical details for the meetings. The facilitator then prepares the session's venue and logistics to ensure that all of the participants' needs for the session are in place.

The workshop starts with the owner defining the objectives, and together with the facilitator, they nominate the participants and agree to a high degree on the session's form. The facilitator usually provides some clarifications and sets up some ground rules.

The facilitator ensures that the session remains focused based on the agenda items, and provides reviews and clarifications where and when required. The task of maintaining focus is a challenging one and the facilitator needs to truly *own* the room and control the session's process. There are multiple techniques to help contain conversations and participants' behavior; and the facilitator needs to be familiar with those invited to the meeting and have a rapport with them. This need for familiarity and rapport are typically one of the main reasons for the BA being the facilitator.

During the meeting, the scribe ensures that outcomes are noted and are visible, and also makes sure the information is available immediately after the workshop. It is important to understand that the scribe's role is not to take the minutes of the meeting, but rather to capture action items and decisions and document them properly.

The facilitator needs to deal with any types of issues or problems that are related to the process, logistics, or flow of the session, as well as conflicts and bottlenecks in the discussion. The role of the facilitator is to ensure that the meeting flows as required, and therefore, he or she needs to be familiar with conflict resolution techniques and how to handle conflicts in such a way that will maximize the degree of resolution of the problem, while minimizing the interruption to the flow of the session. The facilitator also needs to ensure that follow-up items are assigned and looked after.

The responsibility for taking action lies with the participants and the accountability is on the workshop owner, but the facilitator still has an important role in ensuring that action items are followed, along with providing resources and support for those assigned to take action.

The roles and responsibilities in agile workshops are quite similar to those in *regular* workshops—and the same goes for the BA. Beyond having strong facilitation skills, the agile BA needs to keep in mind the concepts around minimalism in documentation and meetings: reduce the duration of the sessions, minimize the number of participants, and always strive to maximize the value that each session adds, while maintaining efficiency.

The Agile BA During the Iteration

The agile BA works closely with the tester during the early stages of the project—and of each iteration—to clarify the acceptance criteria for the iteration. During the iteration, the BA works with the team, as required, to confirm scope and objectives for the iteration and to ensure that the test plan and scripts are in line with the features that are being built. The BA can also help with any additional details for the requirements as they emerge and facilitate the documentation process. The BA also oversees the work with the business as part of the validation process against the acceptance criteria.

Although all team roles are required to participate in the review sessions and demos, the BA may serve as the session's facilitator. At the end of each iteration, a retrospective session takes place—in fact, two sessions, one without the product owner and one with the product owner—where the BA can play an important role in facilitating and ensuring that the action items are properly documented, clarified, and assigned, if necessary. The BA's role in relation to the retrospective session continues after the session because the BA should consolidate the tracking process of the changes proposed and the actions taken as a result of the session. During the next retrospective, the BA will present the findings, report on how the changes were applied and adapted, and whether or to what degree new issues (i.e., side effects) were introduced. In addition to the specific roles mentioned here, the BA should also act as a participant in the retrospective meeting—introducing ideas for improvement of the working process just like any other member of the team.

The Agile BA and Iterative Development

The BA can add value to the development work in multiple ways. While any of the BA's activities mentioned in this book will provide indirect benefits to the development process, there are also a few direct benefits: these include help in defining prototypes, mock-ups, personas, scenarios, and wireframes; facilitating design and technical review sessions; and providing insight based on his or her familiarity with and awareness of the stakeholders' experience and expectations. The BA can also assist in facilitating conversations between the different disciplines within the development team as well as liaising with the release management team and the operations teams in order to ease the transition of the complete solution through implementation.

Kitchen Sink: More Areas of Value Creation

There are additional areas where the BA can add value in agile environments. These areas are not the most obvious ones to think of in the context of business analysis, but rather they are of a broader context, and they can benefit from business analysis skills despite being more generic and organizational-driven. The following list describes these areas:

- *Organizational challenges*: The BA, through facilitation and communication skills, can help set expectations and assist the organization in realizing agile benefits and adopting agile practices, but only for projects that are agile-suitable.
- *Resource challenges*: A BA cannot perform technical work, but having those skills can make it seem as if we have more resources on the project. This is not magic, but rather it is utilizing the business analysis skills to provide answers where possible, or to quickly find a resource, SME, or stakeholder to provide clarifications for escalations and for troubleshooting the problem. A BA can help streamline communication and free up time for the technical resource to continue producing value, while the BA facilitates the pursuit of information to handle bottlenecks and impediments. A BA cannot replace the need for an SME on a project, but can help consolidate information and make the engagement process with the SME more efficient.
- *Agile processes*: Volunteerism is important—and team members need to pick up the stories to work on for each iteration based on a combination of factors that includes priority, complexity and size, their skills, experience, interest, and whether they are the most suitable person to work on a particular task or story. In many environments, the team struggles in this area and the BA can help streamline this process by providing clarification and coordination so that team members pick the most suitable story to work on. The goal is to help turn the team toward becoming self-sufficient, but this may require some hand-holding. The BA can also provide support when the team is not colocated; especially with communication and ground rules.

- *Process challenges*: The BA can step up when there are challenges with the ability of the Scrum Master or the PM to shield the team from external distractions and interruptions. The BA can also help handle challenges that are related to meddling on the part of the stakeholders by helping the team establish rules of engagement with them for sufficient and appropriate levels of involvement. Business analysis skills can also come in handy when the organization attempts to apply some sort of a *hybrid* method that involves only *some* agile concepts.
- *Technology challenges*: As mentioned before, the BA cannot apply any kind of magic to help the project, but he or she can provide coordination and seek help when technical issues hit, including when there is a need to address inadequate testing and deployment automation tools, lack of efficiencies, or lack of focus on building quality code. The BA can clarify issues, articulate problems, facilitate escalations, and perform business analysis and due diligence activities in search of a solution. Additional technical problems that the BA can help with include version controls, knowledge management, and lack of code review standards.

RECAP

Business analysis skills are required in virtually all projects, and as it turns out, even more so in agile projects. As to whether these skills need to be present in the agile project by having them performed by members of the technical team or whether there is a need for a dedicated BA—the jury is still out. With that said, if business analysis skills are utilized properly throughout the project, they can yield a significant set of benefits to the team, through the project, and across the organization.

As business analysis is not part of what the agile *fathers* initially identified, it is viewed to be a newer and emerging concept in agile project management— and naturally, when there is a new role that is introduced into the project, it has an impact (and not always positive) on other team members. In this chapter, we took a look at the role of the BA in agile projects, what is expected of the BA, and the contributions that the BA can make in agile environments. We also took a deeper look into the application of business analysis skills without having a BA, as well as the role impact that the BA can potentially have on other team members and stakeholders throughout the project. We also reviewed a series of activities and challenges where the BA can add value.

It is important to note that we covered multiple areas here and a myriad of activities that the BA can perform in an agile project. With that said, it is not expected that the BA become involved in all of these areas throughout the project. After all, the BA is not a babysitter, and the business analysis skills are there to complement the team's roles and responsibilities—not to replace team members. The contribution of the BA to the agile project should be a factor of the project's needs; the needs and the type of involvement of the product owner; the needs and style of the PM or Scrum Master; the needs, skills, and experience of the team; and the factors surrounding the project—including the environment, organization, and context. There is no one way to have business analysis skills applied to agile projects, but there is an increasing realization that these skills are important to complement the agile team on the way to delivering success.

The somewhat ambiguous nature of the BA role means that the role may be under threat in many agile teams. Organizations that struggle to explain the value of the BA's work may be struggling with benefiting from business analysis skills. At the same time, we need to keep in mind that many of the business analysis skills in agile projects are not very different from what it has always been in *traditional* environments.

Whether similar or different, business analysis skills are important to agile projects and it is important that we find a way to incorporate these skills and maximize the benefit we get from them. Naturally, when we try to do so, there is going to be a role impact (across the project and the organization) and along with it, we should expect adjustments, issues, and even resistance.

ENDNOTES

1. https://www.stephencovey.com/7habits/7habits-habit1.php.
2. *A Guide to the Project Management Body of Knowledge (PMBOK®️ Guide)—Sixth Edition*, Section 4, Trends and Emerging Practices in Project Integration Management, pg. 73.

4

AGILE REQUIREMENTS
AND USER STORIES

THE IMPORTANCE OF REQUIREMENTS

Experience shows that inaccurate requirements gathering has consistently been cited as the primary reason for project failure, and these findings are supported by a Project Management Institute (PMI) *Pulse of the Profession®* study.[1] In fact, PMI's study revealed that close to half (47%) of unsuccessful projects fail to meet goals due to inaccurate requirements management. With such a high number of projects failing due to issues with their requirements process, we need to take a careful look at our requirements management processes and practices and check whether the way we collect, gather, or elicit requirements has something to do with the problems we have with those requirements. Although many organizations lack mature requirements management processes—and some still do not see the value in taking the time to define requirements—overall, there are clear processes, activities, techniques, and approaches to managing requirements. With that said, there might be something deeper and more fundamental with the way we manage requirements.

First, we need to take a look at the terminology that we use—and there has already been significant progress in this area. In the past, we used the terms *collect* or *gather* requirements, but then we shifted toward the word *elicit*. The definition of elicit is: "to draw or bring out or forth; educe; evoke"[2]—which is a much more comprehensive, holistic, and thorough process than just *gathering*. In fact, the word *gather* makes a dangerous assumption: it assumes that the stakeholders we engage and collect information from actually know what they want and need. The danger in this assumption is that often those whom we gather requirements from do not actually know what they need; they may have a concept, an idea, a plan, or a need—but they do not know exactly what they need and cannot articulate that need to us. When we elicit requirements, we take the extra step to engage those stakeholders and get them to first realize their true needs, then articulate it, and finally, communicate it to us so we can write a requirement that fully represents their true needs. With the increase in awareness for the role of the business analyst (BA) in general, the awareness of the proper use of terminology also grew; the word *elicit* is in wide use. Figure 4.1 provides a review of some of the *dangerous* assumptions that we make in *traditional* requirements processes.[3]

Now that we have our definitions in place, it is time to look at the entire requirements management process and check to see if it really serves our needs. In waterfall projects, most of the requirements process takes place early on in the project. A BA is typically in charge of leading and facilitating the process of engaging stakeholders and eliciting requirements. The requirements are then reviewed and subsequently go through the process of baselining, where the project sponsor, customer, and other stakeholders with decision-making capacity

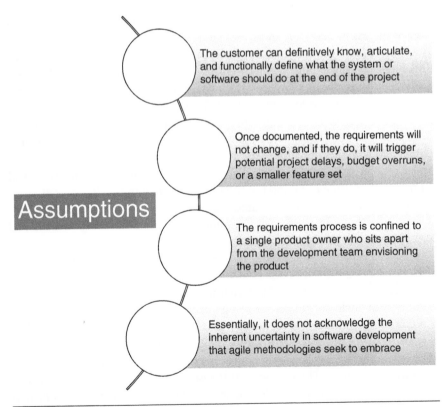

Figure 4.1 *Traditional* requirements

approve the requirements and finalize them. From here, the requirements (that represent the *what*, or the product) are *transferred* to the responsibility of the project manager (PM), where they will be converted into the project scope (the *what*, and in turn, the work breakdown structure that will determine the work to take place to deliver the product).

Although there are techniques and checks and balances built into this process, there are gaps in the hand off of the requirements to the project team and beyond that, the main flaw of this process is that there is an expectation that early on in the project the stakeholders will actually know what they want and will be able to articulate it in such a way that is sufficient for the writing, review, approval, and validation of these requirements. Unfortunately, many stakeholders fail to see or to articulate their true needs at such an early point in the project, and as a result, many of the requirements do not properly, accurately, or fully represent their needs. As the project team starts building the product, new needs will emerge and portions of the product that are being built will be either deemed as the wrong features (building the wrong thing) or as being broken (building the product the wrong way).

Problems with the process around requirements are a result of attempts to save time and money, lack of awareness of the importance of the process, and overconfidence that our stakeholders actually know what they need and that we can capture it as intended. Let's review a few of the common types of conduct that lead to issues and problems with requirements:

- With fixed and aggressive deadlines, organizations often choose to take shortcuts in the requirements definition and they try to design the solution in parallel while working on the requirements.
- As part of an effort to save time (and due to lack of proper planning), the PM and other stakeholders often dictate to the BA a fixed time to define requirements.
- Stakeholders often sign off on requirements without having read them.
- Many organizations schedule long requirements workshops with multiple participants, no clear goal or agenda, and on short notice. The result: many stakeholders cannot attend, the meetings lack focus, they end up missing key contributions and points of view, and the unrealistic expectations lead to further erosion in people's trust in the process. Subsequently, stakeholders do not make themselves available for the endless number of follow ups, clarifications, walkthroughs, and meetings. When the process is unclear, stakeholders end up wondering if there is value in the requirements definition process.
- Many project teams believe that the definition of product requirements is the responsibility of the BA; they attempt to wipe their hands clean of

any involvement in this tedious activity and request to be engaged only when requirements have been finalized.

Product requirements are an important factor on any project—and project requirements, design, building, and testing are all dependent on clear, complete, and accurate product requirements. Many organizations realize this importance, but do not understand how to implement strategies to improve the requirements definition process. With that said, there is an expectation that somehow the BA can produce a clear, correct, and complete set of requirements without changing the flawed practices described here. Overall, the requirements process is a substantial, expensive, and intense process; and when it does not result in good requirements, we need to go through a significant, risky, and costly effort to change the requirements that were previously put together.

The Requirements Management Plan

Since agile methodology does not recognize a distinct role for the BA, many of the work products the BA produces in waterfall projects do not exist, at least formally, in agile projects. However, whether an agile project has a BA or the business analysis skills are performed by team members, there is a need to set up a plan on how the team will perform the requirement (and user story) process: from planning the activities, to assigning roles and responsibilities, and to performing elicitation, analysis, communication, documentation, and management activities.

While the requirements management plan may not be a formal document, there is a need to ensure process effectiveness and keep an eye on efficiencies, looking for the following information:

- *How requirements activities will be planned, tracked, and reported*: This involves defining roles and responsibilities for these activities.
- *Where the requirements will be stored*: Project documentation and file-naming standards need to be consistent across the project and the organization. While we try to minimize documentation, there is still a need to document requirements and their related information beyond backlog maintenance activities.
- *Requirements tracing*: User stories are structured in a way that ensures traceability, but there is still a need to maintain consistency in this area and to ensure that traceability to users and business values is correct. There is also a need to trace requirements beyond what the user story's structure provides—down to design and test activities as well as to project activities. The collaborative nature of agile projects and the involvement

of the team in all stages of the requirements reduce the chance of trace-ability problems, but through the fast pace of events and with many business analysis activities performed by technical team members, some environments may have a need for a BA to oversee the traceability activities. Listing relationships or dependencies between requirements also assists with the requirements allocation process.

- *Requirements attributes*: Here again, there may not be a need for a formal requirements attributes document, but as requirements attributes are used to describe a requirement in greater detail, the team needs to make sure that when the time comes (when the requirement needs to be developed into a feature in the upcoming iteration), all the necessary details about the requirements will be in place. This can be achieved by looking for answers to the following questions that typically need to be asked by the requirements reviewers or by a BA. Table 4.1 helps identify what information we should look for:

  a. Why is this required?
  b. Who came up with this idea in the first place?
  c. Is there another requirement to report on this?

Table 4.1   Requirements attributes (information regarding requirements should be captured and documented with their associated attributes)

Requirement ID	2.1	Unique identifier
Description	The system shall have the ability to track the number of times an application was canceled prior to completion	Requirement description
Rationale	Canceled applications will be used as an indicator to measure the user friendliness of the application	To ensure the requirement provides value
Author	Ori Schibi	Useful when more than 1 BA is on the project
Source	<insert link to meeting notes where this requirement was discussed>	Useful to know who to follow up with if there are additional questions
Complexity	Low	Can be measured by the number of teams required to make changes in order to implement requirement—for example: Low: 1–2 Medium: 3–4 High: 5+

Assumptions	Canceled applications are a result of the application process being too tedious and/or slow	Requirement level assumptions, constraints, dependencies, and risks
Constraints	System constraint—accounts linked to more than 1 canceled application will only be counted once	
Dependencies	Related to requirement 7.5 providing the ability to report on canceled applications	
Risks	Requirements Risk ID 5	
Traced from	Stakeholder requirement 3.2	Backward and forward traceability can be managed using requirements attributes
Traced to Design	To be completed in design	
Traced to Test	To be completed in test	

- *How changes to the requirements will be managed*: Although agile projects are about recognizing and embracing change, the requirements do not actually change in real time. New and emerging needs take place throughout the iteration and new stories are introduced to capture those needs. However, the stories can only be approved by the product owner to be included in the product backlog. Even if a story is introduced in the middle of an iteration, the change will not take place immediately, but rather the story will be prioritized (ultimately by the product owner) and determined whether it is going to be performed in the next iteration (and then it is estimated by the team and broken down into details and tasks) or in a future one. The iteration review (or demo) at the end of the iteration will also allow changes to be introduced—whether due to performance (defects) or due to new and changing needs as a result of the demo. Either way, every time a story that was demonstrated is not accepted as-is, a new story will be introduced and prioritized accordingly. The notion of changes being randomly introduced and worked on does not exist in agile projects. However, the prioritization, planning, and estimating processes of new and changing stories (as part of backlog maintenance) should be planned for and everyone needs to understand what is expected of them to ensure the process is followed and that it yields a realistic backlog that addresses the customer's needs. Chapters 6 and 7 discuss in detail the estimating and the backlog maintenance processes in agile projects, along with handling change and prioritization. The fast pace, the closer working relations between the team and the product owner, and the shared responsibility in identifying and working on new stories do not imply that the change process in agile projects does not exist. What was described here is a change management process that ensures full accountability, prioritization, capacity management, and impact assessment of the change. It is the product owner who makes the final call and provides a formal sign off for all changes that are to be incorporated in the upcoming iteration. All the due diligence is taking place between the iterations, where all proposed changes are considered, all impacted stakeholders are engaged, and approvals are provided. To save time and cost, reduce documentation, reduce risk, and ensure process efficiency and effectiveness, at the end of each iteration changes are submitted as one *omnibus* change request for the approval process.

• *How requirements will be prioritized*: This task is a fundamental ingredient of agile project success. While it is the responsibility of the product owner to prioritize user stories and requirements, the process is done through collaboration and engagement of team members, subject matter experts (SMEs), and other stakeholders. The BA can set up a plan to ensure that expectations are clear, along with timing, resource needs, and decision-making capacity—so all those involved in the process are on board.

Product versus Project Requirements

Before moving on with an overview of agile requirements, there is one more thing to clarify: the difference between *product* requirements and *project* requirements. Product requirements are defined based on the business requirements and the solution scope defined prior to the project as part of the organizational needs and business case. Building product requirements on top of a weak foundation leads to delivering a product or service that does not meet the needs of the stakeholders. Product requirements (what we need to build) describe the goods or services that the project will deliver, while the project requirements (how we are going to build the product) describe the work or activities that are required in order to deliver the product. There are dependencies between product and project requirements and the two must be developed in conjunction with each other. To be clear—the project requirements are really the ones that depend on the product requirements; and the requirements management process we discuss here refers to the product requirements. Any attempts to change product requirements must be subsequently considered and accounted for as part of the project requirements. The product requirements process engages the business stakeholders early on to identify needs and constraints, and it is followed by the project requirements that are put together by engaging the core project team (PM, technical leads, and test leads) based on the product requirements. Table 4.2 shows a simplified view of the difference between the product requirements and the project requirements.

Table 4.2　Kitchen renovation

Product Requirements	Project Requirements
Type of appliances, lighting, materials— countertops (granite, Caesarstone, soap stone) size and measurements, colors, and materials of cabinets	Resources required to install the kitchen— plumber, electrician, carpenter, time to complete work, budget, and all activities to perform (work itself)
BA accountable	PM accountable
BA to elicit product requirements within project constraints	PM cannot determine all project requirements unless product requirements are known

This table simplifies the difference between the product requirements and the project requirements. Product requirements reflect the product (what we are going to build), while project requirements are about the resources and the work that are involved in building the product.

OVERVIEW OF AGILE REQUIREMENTS

Agile methodologies recognize the fact that we do not know everything and that we are always learning, therefore it is not realistic to expect that stakeholders know all of their needs up front. As a result, agile methodologies call for eliciting high-level requirements early in the project and then dive deeper only for the requirements that are due to be built into features in the upcoming iteration.

Agile methodologies focus on implementing small pieces of the project's product at a time. The agile teams are dedicated, focused teams that engage in a collaborative process to release functionality into production in cycles (i.e., sprints or iterations) that typically last between one and four weeks.

While the different layers of requirements are not as apparent, the process is iterative in nature:

1. We start with eliciting high-level product requirements (user stories) early in the project, along with high-level understanding of the project requirements
2. We will then provide high-level estimates (through sizing the requirements and determining their complexity by determining their story points, or ideal days)
3. As part of the planning and estimating process, we will refine some (around 6–8) of the high-level requirements into a detailed level (determining the project requirements) and then provide detailed estimates to allow consistency of the sizing exercise and the mapping of these requirements to determine their durations
4. The requirements will be distributed into iterations and releases (based on business needs, prioritization, capacity, and team velocity) and for each upcoming iteration, the team will break down the high-level

requirements for the iteration into detailed requirements (involving project requirements—all the work, tasks, and activities that need to take place in order to produce the product increment), and subsequently the team will perform detailed estimates for these requirements

User Stories and Use Cases

Agile requirements are written in the form of a user story, which is a high-level description of system behavior with a specific structure: As a <user>, I want to <perform functionality>, so that <I can realize business value>. This format ensures that the user story captures some very important items in addition to functionality. These include the traceability to the user (to know which stakeholder is behind the need and the requirement) as well as traceability to business value (to ensure that whatever we build here adds value for the customer). This means that stakeholder requirements are captured as part of the user story, along with the solution requirements. The functional requirements will then be combined into use cases and/or user stories, which in turn, will provide insight into non-functional requirements. Figure 4.2 illustrates the relationship between user stories.

User stories tend to be *lightweight* specifications and requirements. In fact, user stories are not a full specification of the requirement, but a placeholder for conversation about the requirement when more information is needed. The user story will be fully specified and clarified upon being brought into the iteration and the development cycle. Once delivered, the user story represents a fully functional slice of the overall system.

BAs commonly utilize use cases in support of user stories and to substantiate them or they incorporate use cases as a source of business requirements that eventually lead to user stories. A detailed discussion of use cases takes place

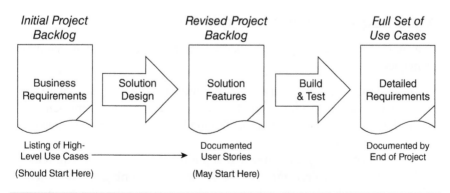

Figure 4.2   User stories and use cases

" As **an owner of multiple stores**,
I want to **motivate my employees**
so that **my store will meet profit goals.** *"*

Use Cases: Goal = **View Dashboard**
Actor = Owner of Multiple Stores
Context = Remotely on Tablet
Scenario = Check sales target to see if store has met sales goal

• Actions = View sales goals for multiple locations   • System Activity = Allow access to data for multiple stores
based on single log in

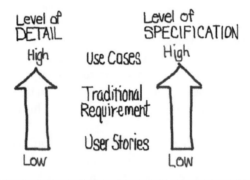

Figure 4.3   User stories and use cases are not the same

later in this chapter. User stories are not the same as use cases, but both serve a
similar purpose: to help understand what the user needs. Both are augmented
with models to specify expected behavior. Figure 4.3[4, 5] highlights the differences between user stories and use cases.

Agile Requirements and Agile Principles

The agile requirements process focuses on addressing agile development
principles:

- *Improve visibility*: The user story is written in such a way so that we know
  whose needs we address and the value the requirements represent.
- *Improve customer satisfaction*: Tracing back to the user's true needs ensures that we address those needs.
- *Improve project flexibility*: We first put together high-level requirements
  early in the project, and then we go through a refining process when we
  need more details in order to build the features that the requirement
  represents.

- *Improve responsiveness for change*: The high-level requirements represent our understanding of the needs and we are open to changing the requirements as needed based on progress, new needs, changing needs and conditions, and the better level of understanding by the customer that may represent an evolution of the needs.
- *Improve delivery of value to customer*: If the requirement does not show that it represents value for the customer, it should not be done. The role breakdown in agile projects ensures that the needs are defined by the business (specifically, the product owner) and the *how* (the work to be done) is then defined by the team. Technical members of the team do not identify needs or features and the approval process ensures that any idea that is brought up has to be approved by the product owner to ensure realization of value.
- *Reduce time to delivery*: For some it may seem ironic that by having ambiguous and high-level requirements, we actually focus on delivering early value. However, we can deliver early value by not wasting our time finalizing requirements that may change or get canceled later altogether. We focus only on value-added activities which means just enough information just in time. It also helps ensure that we rely on the latest information to build a product increment that represents a customer need that can be released for customer use.
- *Reduce risk*: We employ all of the aforementioned techniques, along with graphical, visual, and constant accurate reporting to ensure that we build the right thing and that we build it the right way.

User stories introduce many benefits for the customer and the agile team: they help instill a good customer relationship; they are structured in a useful way for iterative and incremental delivery (or agile projects); when stakeholders and delivery teams are colocated and work well collaboratively, user stories can help reduce the need for more detailed requirements specifications; they clearly define the value the solution provides; and they serve as a strong foundation for building realistic estimates.

However, user stories also introduce some challenges that the team needs to consider and address—sometimes with the help of a BA: story formats may not include sufficient details when there are challenges around collaboration or when team members are not colocated; in their early, high-level form, they are not detailed enough when there is a greater need for details (i.e., in regular environments); user stories are not aligned with waterfall methods that may take place in the organization; user stories may introduce challenges in documentation-centric environments; and finally, user stories typically do not document non-functional requirements effectively, therefore, they require additional techniques and modeling options to ensure full coverage of the requirements.

THE DIFFERENCE BETWEEN AGILE AND *TRADITIONAL* REQUIREMENTS

While there are differences between agile and *traditional* requirements, we need to keep in mind two things:

1. In essence, there are many elements from *traditional* requirements that need to take place when working on requirements in agile projects
2. Similar to many other things, we first need to know the basics and foundations of how to do things (that is, how to perform traditional requirements) before we move on to a more advanced way of doing things

Agile projects have many moving parts; things take place at a faster pace and through short and quite intense cycles. The reference to the *traditional* way of doing requirements focuses on waterfall, or linear projects, where there is a distinct *phase* of collecting, gathering, and eliciting requirements. At the end of the phase, we typically baseline the requirements, which is the process of getting the requirements reviewed and approved by the appropriate stakeholders (typically the sponsor, customer, and other decision makers). Having such a distinct *phase* in which to do the requirements makes it easier for those who work on it because their effort is condensed and focused on working on the requirements through a specific period of time. Table 4.3 provides a summary of the key differences between agile and traditional requirements.

There are some similarities between agile and waterfall requirements, including the need to figure out the details for the requirements before getting the team to perform the work and build the features that are associated with the requirements; however, the timing is different. In agile, we will get to the detailed requirements only when we need to do the work, while in waterfall that is traditionally done early on and is typically baselined before any development work is taking place. Both types of requirements need collaboration and team engagement, but in agile requirements the focus is on group discussion and consensus, while in waterfall requirements the focus is on formal documentation and evaluation.

Changes to Requirements

Although changes to requirements are expected, agile projects expect changes to occur and even embrace change. The entire process supports the prospect of changing requirements—from the high-level requirements and planning during the release planning to the breakdown of the project into shorter, more manageable iterations. Any new or changed requirements are added to the product backlog at any time, while keeping iteration goals untouched (except for significant *game changers*). Changes need to be approved by the product owner. An

Table 4.3   Agile versus traditional requirements

Considerations	Agile Requirements	Traditional Requirements
Timing	High-level early, and then short pulses for each iteration	A large requirements phase at the beginning. Adjustments as we go
Clarity	Ambiguity is allowed until further information is needed. Course-grain requirements	An attempt to finalize and baseline the detailed requirements up front
Structure	A user story is written for the user and contains a reference for the user and the business benefits, in addition to the functionality	Written from the solution or the system's point of view. Does not have traceability built-in to the requirements and with a solution-centric language, there is a greater chance that the requirement does not address a true need or is not expressed correctly
Expectations	We know that we don't know everything, and we come to terms with the fact that we keep learning as we go	We "pretend" that what the customer just told us is what we need to build
Focus	The customer (and the user)	Those who do the process and the solution
Changes and new needs	Designed to accommodate changing needs (new needs, changes to expressed needs, and new realization of value)	Once the requirements are baselined, changes are an after-thought that is reactive by nature. As such, the process becomes pricier and more risky for the project
Overall approach	Rolling wave planning	Patches and adjustments
Done	We clearly define success criteria to ensure we build the right thing and in the right way	Acceptance criteria is often not defined properly or sufficiently

organization can abort an iteration at any time to refocus the team on a new higher priority set of goals that have emerged. Product backlog is intentionally dynamic and there are no frozen requirements. Prioritization is a critical part of enabling the effective process of managing requirements (and the entire agile project)—with specific and enhanced importance when we deal with changes. When a story is being added to the backlog, it is not enough to just assess, review, and approve the change—it is paramount to prioritize it into the backlog and determine where it fits (before which of the other requirements). As a result of a change, it becomes easier to articulate the impact and the trade-offs that the change introduces because we can clearly demonstrate which requirement will be *bumped* as a result of accepting the requirement as part of the change.

Waterfall projects also go through changing requirements, but the only mechanisms available are part of the change control process. When a change is introduced, it needs to follow that process (which is often heavy-handed, long, and cumbersome). The process involves an assessment and review of the projected change impact, and if a change is accepted or approved, it will be incorporated into a new document version of the requirements. The process, by nature, is reactive, but even if it is followed, with the lack of a clear prioritization process it is hard to determine where exactly the new requirement fits in and what the exact impact will be on other requirements. This makes it harder to properly price the impact of the change or to determine where the requirement will fit in. Another common danger is that many teams do not have or do not follow a proper change control process—allowing changes to be introduced, assessed, approved, and worked without proper impact assessment, pricing or capacity assessment, or prioritization. This can result in a failure to meet business needs, removal of important requirements, resource allocation issues, unmanageable workloads, failures due to time and cost overruns, or quality issues. When many changes take place in a waterfall project, it quickly overwhelms the team and creates chaos due to the multiple changes, random order, lack of prioritization, and failure to realize the team's capacity. Even the contribution of a dedicated BA may not help overcome these challenges. One of the main benefits that agile approaches offer is a more proactive and responsible way to handle the changing needs throughout the project.

Whose Need Do We Address?

With a clear and distinct requirements elicitation and analysis *phase*, stakeholders and team members develop a strong awareness of the process; people's state of mind is typically focused on doing the requirements and once this is done, the focus shifts toward ensuring that the requirements are mapped against the project scope before starting to build the product. This is done with the intent of making the process clear and easy to perform for those who do the work. However, when we design a process, there is another element to consider: whose needs should we be focusing on—the needs of those who do the work or the needs of the user and the customer? Making the process *easier* and straightforward toward the needs of those performing it may not yield the best possible results, since that usually means that the process is not designed toward that of the *customer* or the end user. It actually implies that those whom the process was designed for were the ones the process focuses on, while in this case, the customer and the end user may be a mere afterthought.

For example, when we need to call our favorite telecom provider for two matters (e.g., technical support and a billing issue), we will need to wait in two different queues—one for the technical support, and once they resolve our technical issue, they transfer us to the end of the line in order to wait in the queue for the billing department. One call, two needs, two different rounds of waiting and dealing with two different representatives. The reason for this convoluted series of events is that the process is designed for the convenience of the service provider with the focus on training and cost consideration and with little or no regard to the needs of the customer. This does not imply that it is easy to design streamlined processes that allow the customer to have all their needs addressed effectively through one customer service representative, but because of constraints related to costs, organizational structure, and process mapping, the customer, in many cases, is an afterthought rather than the focus. Thus, when the requirements process is designed for those who work on the requirements process, it is more than likely that the process is not designed with the *users* of the requirements in mind.

The agile requirements process is designed with more focus on the customer. It is important to have a clear, effective, and efficient process for the team to perform the requirements process, but the needs of those doing the work are only secondary to the needs of the customer. Focusing on the customer's needs means that it is more likely that the agile requirements process will yield the correct requirements that fully reflect the customer's needs.

Another major difference is that in agile projects there is no expectation that the customer and other stakeholders know exactly what they want from the onset. In traditional projects we assume that the customer knows what they want and we give stakeholders a false sense of confidence that we understand those needs and that we properly captured those needs in the requirements. We refer to it here as a false sense of security because it gives the impression that we are all *on the same page*, but no one really knows what page it is, or even which book we're reading. Even if we properly capture what the customer told us, this does not mean that the customer actually has a full understanding of his or her own needs. We then take this confidence and start building a product that may not be the right product, or we may be building it the wrong way.

Agile projects focus on reducing risk—and specifically the two risks that were mentioned here: (1) we reduce the risk of building the wrong thing (enduring conformance to requirements) and (2) we reduce the risk of building the right thing, the wrong way (reducing the chance we will be producing defects and ensuring that the product is fit for use).

DEVELOPING USER STORIES AND REQUIREMENTS

A user story describes what a user wants to do. It is a structured narrative that provides value to the user, has specific acceptance criteria, and documents the requirements with the user's point of view—focusing on the benefits of the functionality requested. A user story serves as a placeholder for a requirement about which the team will have a conversation with the product owner during an iteration. The team, with the help of the BA, needs to discover the details of the product owner's needs.

The three-part structure of the user story focuses on the user/actor who represents the role requiring the functionality; the functionality—to describe what the user needs to be able to do; and finally, the benefits—representing the value the requirement will bring. The agile team needs to develop user stories by engaging stakeholders and eliciting information—exercising their business analysis skills. When there is a team member who acts as a BA, he or she facilitates the conversation between stakeholders and developers to get an understanding of the what/why/who of the requirement in simple and concise terms. While user stories are very useful in helping developers understand user needs in a more effective way than formal requirements, some requirements need to be expressed in the formal (waterfall-like) way. For example, solutions and systems in a regulated industry must be compliant with the regulations and there is no point in trying to force these requirements into the structure of a user story. It is important, though, to ensure everyone's understanding of the acceptance criteria, the constraints, and the related non-functional requirements. At times, non-functional requirements can be incorporated into the acceptance criteria of the solution requirements.

During conversations, the goal is to try to understand the intent of the user story. The team, with support from the BA, should engage the product owner and the SMEs and ask them questions that will help obtain the necessary information. In a way, seeking the information in the following list can be interpreted as the application of use-case principles to the development and elaboration of the user story:

- Identify possible scenarios and describe the details of each scenario
- Define each scenario's preconditions to understand what is assumed as truth in order for the feature that we build to work as expected
- Define each scenario's post conditions so we know what needs to be the result of what we build when the goal is accomplished
- The BA needs to challenge the conversation and ask for each scenario step, whether or not it can be done in a different way
- Finally, we need to check if every user acts the same way. If different users utilize the solution in different ways, there is a need to define personas (discussed later in this chapter) to understand the different user classes

We Need the BA

Members of the agile team participate in the requirements process and, ideally, virtually all team members need to demonstrate business analysis skills to ensure a proper requirements process all around—elicitation, analysis, documentation, communication, and the validation process. While it is realistic to expect to have team members who can demonstrate business analysis skills by *wearing two hats*, there are better chances to achieve it in smaller projects and in small organizations. With the growth of agile adaptation across industries and the increase in the number of agile projects, it may become more challenging to find team members who can easily perform both their technical role, as well as demonstrate business analysis skills. When resources need to perform the dual role, it is likely that business analysis will be their secondary skill, and therefore it will be pushed aside whenever there is any conflict over the split of their time between the different skills. The challenges grow even bigger when we need to figure out who is going to perform specific business analysis actions.

When there are challenges in getting resources to perform the dual roles of technical work and business analysis, it is time to consider getting a BA. This means having a person whose title is a BA, rather than having the business analysis skills being an afterthought. With a BA, it gets easier to identify specific business analysis tasks and gain the confidence that the BA will bring to perform these tasks properly and in a timely fashion. For example, it is the BA's responsibility to communicate requirements and ensure the information about requirements flows; it may not be sufficient to just leave this task to the team members to perform since requirements communication needs to involve many people, including SMEs, business stakeholders, and technical team members. Having a BA as part of the team does not preclude team members from having to communicate requirements, but it provides some piece of mind that the BA at least oversees the process and covers any gaps in the requirements communication. Requirements communication should include constraints and goals and it can be done through conversations (verbally) and documentation (written).

This brings us to another requirements-related task to perform: documentation. There is more responsibility on the team to perform the documentation than there is with communication. While it is expected that the BA *owns* the requirements communication, it is the responsibility of the team members, and not that of the BA, to perform requirements documentation. The BA can help facilitate the process, but it will be a waste of resources and time if the BA actually needs to document the requirements, instead of the team doing so.

The BA and the Product Owner

The development team and the product owner must have a common vision—and this is accomplished by having the entire team work with the product owner. The product owner must be present to answer questions from the development team during the project. If this process does not work well—whether due to the availability of the product owner and the SMEs or because of team-related issues (i.e., availability, allocation, communication skills)—the BA is expected to bridge the gap by looking for additional documentation, clarifying inquiries, consolidating information, or engaging additional stakeholders as needed. These actions by the BA help reduce the risk on the project because of any breakdowns in communication between the team and the product owner and other stakeholders.

When a product owner or the user is not available, a *proxy* can take their place: a user proxy is an authorized person who can be relied upon to speak or act on behalf of an actual user. While the proxy is not authorized to make decisions beyond the mandate provided by the user or the product owner, he or she can help streamline communication, provide clarification, and reduce redundancies in communication. When there are challenges around the product owner's or the user's availability, the selection of the correct proxy is critical to the project's success since a proxy with the proper domain expertise can successfully represent the user on an agile project. In many situations, the BA can act as a proxy; fulfilling the bridging role that the BA brings to the table.

Back to Developing Requirements

When developing user stories, smaller stories with a clearly defined goal are better, and we should try to have stories that are small enough so they can be implementable in a single iteration. Since stories need to be written first at a high level—with details omitted until implementation starts and we get to dig deeper and add details to the story—that does not mean that the story gets bigger, but it means the breakdown of the story into smaller, more manageable pieces. We need to make sure that the process of getting more details for stories does not lead to scope creep, feature creep, or story creep—all referring to an increase in the coverage of the story, instead of an increase in the level of details within the mandated boundaries of the story.

During the development of a story, anyone can contribute to it (i.e., the product owner, users, analysts, SMEs, developers, testers), but at the end it is up to the product owner to approve the story. The role of the BA in putting together requirements in agile environments is to facilitate the iterative process and ensure the development of detailed stories by taking the high-level and ambiguous requirements we put together early on and get them elaborated and refined into a sufficient level of detail at the right time.

Writing User Stories

Although the process of writing and documenting requirements is significantly lighter in agile projects (as compared to waterfall), there is a need to ensure that the stories are properly written. The fast pace, ambiguity, less documentation, less formality in general, and the approach of *just enough* and *just in time* should not mislead agile teams into thinking that there is no need to follow processes—or that there is a need to rush through activities without producing results. Specifically for requirements and story writing, there is a strong need to do it properly so that we can build *good* requirements that the team can work with and build the product around. A discussion on what makes a good requirement takes place later in this chapter (Bill Wake's INVEST criteria).

A common approach to creating the initial set of user stories is a story writing workshop. The team, along with stakeholders and SMEs, need to work together in a facilitated session that is similar by nature to *traditional* requirements sessions, commonly known as a joint application design (JAD) session.[6] The JAD is a process used to collect business requirements while developing a new product or solution. The process involves techniques to enhance user participation, expedite development, and improve the quality of specifications. While workshops are expensive (several hours in total, multiple participants), hard to lead (requires facilitation and involves multiple points of views, styles, and goals), and difficult to schedule (logistics and participants' availability), it is one of the most effective ways to gather, elicit, and capture requirements. The workshop helps produce the high-level stories early in the project; and additional requirements will be discovered by the team (along with the BA) during iteration implementation and by stakeholders and the product owner during iteration reviews. Stories that are captured do not need to be discussed in detail yet until further information is needed for them.

Despite the workshop being less structured than a JAD session, there is a need to establish ground rules and expectations to ensure session success. It starts by ensuring that everyone who is committed to the project participates, including mandatory attendance by (but not limited to) the product owner, developers, testers, and the BA. The session should be facilitated (by the team manager, Scrum Master, or even the BA). Participants must have (or obtain through the session) a clear understanding of the product vision and they should try to refrain from critiquing stories during the elicitation process. This session should be time-boxed (try to keep it under two hours)—and it is not the time to try to break epics (discussed later). We need to ensure that the expectation is that not all stories will be captured during this workshop.

During the session, there is a need to ensure that user stories are brainstormed for each user role and that all functional areas are considered (e.g., development, testing). A simple, low-fidelity throwaway prototype can help users visualize what they are looking for and we should write a story for each needed feature while separately capturing and documenting constraints.

Risk Clause

The user story format can be extended to include a *risk* clause if required. This can help provide stakeholders with context, as well as with an understanding of the stakes involved if we fail to achieve the goal of this user story. For example,

"As a hiring manager, I want to post an open position, so that I can find qualified applicants for interviewing. If I cannot find candidates for the position, it will remain unfilled or it will be filled with an unqualified candidate."

User Stories, Epics, and Themes

User stories, epics, and themes are the main ways to express needs and develop requirements in agile projects and although all three have the same format, they differ from each other—with each having a purpose—and sometimes those differences lead to confusion. There are additional formats in which to express requirements; and the team, along with the BA, needs to determine which formats should be utilized for each requirement to achieve effectiveness and efficiency of the process:

- *Features*: something the product has to do or to be; it is commonly used to express non-functional requirements. A feature can have the following format: When <an initial condition>, the product shall <action>. The format of a feature resembles a *traditional* requirement of the type we are used to seeing in waterfall projects; but in the context of agile projects, features are used at a higher level than detailed requirements and user stories will be developed based on these features. In some cases, this format can also be used for developer stories or to describe non-functional requirements.
- *Scenarios and use cases*: step-by-step descriptions of a user action, or a set of such actions; these are diagrams that can be supported by a narrative or may simply be just a narrative (covered in detail later in this chapter).
- *Diagram and visual models*: can also be used to capture requirements since they are useful to help communicate ideas effectively and relatively quickly. At times, the visuals can serve as a precursor to developing user stories, or as requirements.

Epics

When we reach an iteration in which a story needs to be developed, the story has to be small enough to fit into the iteration. When the process of breaking down a story does not yield a small enough story that can be developed in one increment, it is called an epic. The characteristics of an epic are similar to those of a *regular* story, but the epic is big. In other words, the epic is a description of what a user wants to do, it provides value to the user, and it has specific acceptance criteria. While the epic has to be split into true user stories, each with value to the user, the epic cannot be split only into true user stories.

There are two common views of what is an epic though they contradict each other:

1. An epic story must be split into true user stories each with value to the user. Although the entire value of the epic cannot be achieved through one iteration, the split stories should each fit into an iteration. Typical ways to break epics down include workflow steps, different types of data, different business rules, different versions, or create, read, update, delete (CRUD). Each of the smaller stories should have its own acceptance criteria. This is a more common way to refer to epics.

2. The second way to refer to an epic is that it cannot be split only into true user stories, but it can be broken down into *developer stories*. These types of stories do not provide value to the original user on their own despite each having acceptance criteria.

Despite the definitions contradicting each other, in practice, there is not much difference between them as they both break down the epic into smaller stories. The difference is in the semantics regarding whether the smaller stories are named user stories or developer stories and the user on whose behalf we write the story. Therefore, we can say that an epic is any user story that is too big to fit into a single iteration and so it must be split. The following is an example of an epic that is split into smaller developer stories: As a regional salesperson, I need to engage clients across the entire region. Due to the different regions, time zones, languages, and legal requirements from one country to another, this story will take more than one iteration to implement. A series of smaller stories (in this case, developer stories) will need to be developed; and as developer stories, their wording will be different than what we use in *regular* stories. Developer stories will begin with "As a developer," or "As a product owner," to prevent confusion and to clarify that these stories do not provide direct value for the user despite each being important to the process of completing the user story they pertain to.

The BA can help the team articulate a way to treat epics and criteria to determine when an epic cannot be broken down further.

Themes

A theme is a collection of user stories or epics that are useful to the team. Themes are used to organize stories into releases—and many themes inspire user stories. Themes can be organized around user types (e.g., for a service department in a car dealership: customer service representative, administration, mechanic, customers), around organizational units (e.g., for a car dealership: sales, body shop, service), or around different large scenarios (e.g., leasing, financing, service, recalls). The structure depends on how the users see the work. Themes are often used to organize stories into releases (and iterations) or given to sub-teams to work on. Themes can also be there to inspire user stories. The BA can help the team ensure consistency in the process of identifying and, in turn, breaking down themes into stories.

The process of determining that a story resembles an epic is important because there is a need to ensure that information flows and that progress is measured and reported. There is also a need to prioritize, reprioritize, articulate the themes and stories—and to determine what will be included in each iteration. While there are clear roles and responsibilities in agile projects, it is the responsibility of the team (and respectively, the product owner) to do these things mentioned here—but a BA can become critical in closing gaps, reducing duplication of effort, introducing consistency, and streamlining these processes. Any challenges the team faces like breakdowns in communication, misunderstandings, or a product owner who is not sufficiently available and involved may lead to problems in putting stories together and to working with epics and themes.

INVEST in a Good Story

During the release planning the team works with the product owner and with other stakeholders to identify stories. These stories are defined only at a high level and some of these stories will actually be themes (for releases or iterations, and they will be broken down later) or epics (that will be performed over more than one future iteration). The requirements details need to be figured out as late as possible, allowing ambiguity until further information is needed. The details of the stories need to be determined close to when the team needs to start working on the stories and developing them into features and functionalities. Therefore, the word *later* is not a call for procrastination, but rather *later* means that we understand the solution and the big picture better, we understand better how each requirement fits in, there are less unknowns since uncertainties are resolved with time, and there is less effort and time (waste) toward early understanding of requirements that will end up changing (or even being canceled) later.

With that said, we still need to keep in mind that there is a need for high-level requirements early in the project and because of that we need to prepare the overall solution outline and model. Although the BA is not a technical authority, he or she can be instrumental in the early definition and communication of the high-level requirements as well as in the cyclical process of breaking down the requirements and figuring out their details.

The process of putting together the detailed requirements is not easy. A lot of information is flowing in from multiple participants, stakeholders, and SMEs— and this information may go through several team members who represent different roles and functional areas. There is the potential of information getting lost, being misrepresented, or being misinterpreted; and since there are several requirements that go through the detailing process in each iteration, the BA can help ensure there are no issues with the process or the information flow.

There is also a need to define what constitutes a good requirement; that is—a requirement that is in a state that is ready for the technical team to work with it and build the product. As agile methodologies are big on defining acceptance criteria and defining how to measure whether the product that we build is truly *done*, here too, there is a mechanism to ensure that each requirement is *good*.

Bill Wake introduced a set of characteristics that make a good requirement[7] (and a good backlog), saying that you need to INVEST in a good requirement where the word *INVEST* stands for the following:

- *Independent*: The story needs to be self contained so there is no inherent dependency on another story in the backlog. Avoid dependencies with other stories.
- *Negotiable*: A story is not a contract and it should be written in such a way that it leaves room for further discussion. Too much detail up front gives the impression that more discussion on the story is not necessary. We should keep in mind that not every story must be negotiable (e.g., constraints are not negotiable).
- *Valuable*: A story must represent value for the customer, the user, and other stakeholders.
- *Estimable*: Stories must be written in such a way so that we can estimate their size. The team will encounter problems estimating if the story is too big, if insufficient information is provided, or if there is a lack of domain knowledge.
- *Small*: A story has to be small enough so that it can be possible to plan, estimate, task, and prioritize it with a sufficient level of accuracy. The word *small* also refers to the story size because it should be small enough to be fit into and to be completed in a single iteration. Stories that need to be worked on in the near future should be smaller and more detailed; and larger stories are acceptable if planned further out. Epics are larger stories that do not fit into a single iteration.
- *Testable*: The story, or any other information that supports it, needs to have sufficient information to have it tested. Acceptance criteria should be stated in customer terms and be clear to all team members.

The INVEST criteria can be seen with short explanations in Figure 4.4. It is very useful in determining whether or not a story is good, but in many situations and for many requirements, it may be challenging to meet the criteria of the definitions for each item within INVEST. It is therefore important not only to familiarize the team with the INVEST criteria, but also to ensure that the team has the means to meet that criteria. Whenever there is a challenge with meeting the criteria, the BA can help facilitate the process of articulating and meeting the INVEST criteria or pursuing the relevant information for each story on the backlog.

> **Independent**
>
> • Keep stories independent from one another as much as possible

> **Negotiable**
>
> • Story details are not fixed and can be worked out as needed

> **Valuable**
>
> • Story has value to the customer

> **Estimable**
>
> • Stories are written in such a way that allows developers to "size" them

> **Small**
>
> • The story can be completed within a single iteration

> **Testable**
>
> • Story must be testable

Figure 4.4   Invest in a good story

PERSONAS

A persona is a description of a fictitious user based on assumptions and data collected through research and information gathering. Personas represent different users within a single user role and this method can be beneficial in helping identify stories in user interface design. It can also help the team and the designers consider different perspectives that different users may have, rather than only consider their own point of view.

For example, there can be three types of data entry clerks in an organization:

1. *Novice clerks*—who know their role but do not know how to perform complex activities or how to fix any issue they encounter
2. *Typical clerks*—who know how to handle the most common challenges that they may encounter
3. *Super clerks*—who know how to troubleshoot any type of situation and constantly strive to challenge and improve their work environment

A persona is not a description of a specific individual, but rather is a partial description of the aspects relevant for the design effort of the solution. Personas generally include a series of information that will be useful for the designers and the team in the process of building the solution, as described in Table 4.4. There are many benefits in using the persona method:

Table 4.4 Personas information—this information will help the designers and the team identify the different types of users

Persona Information	Details
Personal information and demographics	Name, gender, age, and picture
Background	Education, experience, job title, income level
Likes and dislikes	Additional information about preferences, style, and behavioral information
Approach to work	How this persona handles situations, problem solving, and reaction to problems
Goals	Needs, objectives, motivators, and ambitions

- Helps better understand the product's users, customers, and stakeholders
- Speeds up the design cycle by allowing more focus
- Improves the chance of addressing true users' needs
- Helps produce a more usable solution that fits the actual needs

With the benefits in mind, we need to remember that the persona method has not been fully defined and is utilized differently in different environments. The personas must be carefully crafted to ensure that they truly represent the user space accurately; and there is a danger that those who are involved may get *carried away* by over-focusing on demographics or on other information that may sidetrack the effort of learning about the needs of the users. The persona needs to remind the designers of someone they know, but a persona is not supposed to represent a specific individual.

The process of identifying personas involves performing a variety of information gathering techniques, including market research, surveys, focus groups, and interviews. Once the personas have been identified, the team should move to defining the user stories that describe how each persona is most likely going to use the product. However, we need to keep in mind that we may not be able to address the needs of all personas and therefore we need to prioritize the personas. By prioritizing the personas, we will be able to ensure that we address the essential needs of the personas that we identified as more important. This will ensure that those needs are addressed earlier and that when we need to reduce scope, the less important needs will be the ones who do not make it. The prioritization process involves a series of issues to be considered—listed here in no particular order:

- Anticipated number of instances of a persona
- Consider the influence of persona instances
- Ensure alignment with core competencies

- Measure the total usage and frequency of use
- Analyze the market potential, competitiveness, and opportunities

The information about each persona and the prioritization process are in place to help the team and product owners *pick their battles* and determine which needs, of which personas, need to be met as a higher priority—knowing that it is likely that not all needs of all personas will be addressed. Due to the different needs represented by personas, there may be a need for different products or designs for different users. Starting with the most important personas, the agile team will address the needs of different users through different releases. It is the job of the team and the designers to come up with the personas and to identify the needs of the various users, but a BA can help maintain focus on valid information, as well as support the prioritization process—always keeping the business value in mind. The BA can also assist the team in asking the right questions throughout the process of applying the personas and working on the requirements. These questions can help address disagreements and misunderstandings in relation to the scope, the design, or the actual needs:

- Would <the user> use this?
- Would this feature be valuable for <the user>?
- What does <the user> need in order to achieve his/her goals?
- Does this only apply to <this user> or to others as well?

The process of identifying personas is illustrated in Figure 4.5.

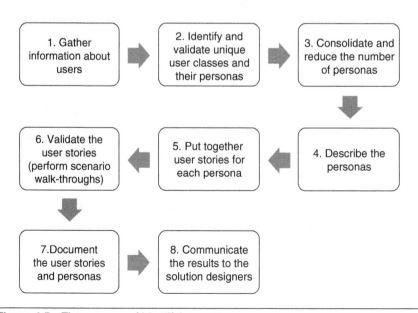

Figure 4.5   The process of identifying personas

USE CASES

Earlier in the chapter we touched on use cases and their role in either articulating business requirements or providing support to the process of building user stories and requirements. Let's take a more detailed look at the characteristics, the benefits, and the process of building use cases.

A use case is an interaction that a user (known as an *actor*) has with a system or a solution. It is a process-modeling technique that focuses on areas of the business that have been selected for automation. Most actors are human users, but they can also be external systems that interact with our solution. Each use case represents a goal that the actor needs to achieve (e.g., a student who tries to enroll in a university course by using an online enrollment tool). A use case represents a single goal for a single actor.

Use cases and user stories support each other—but they are not the same and are not to be confused with each other. User stories are centered on the result and the benefit of what we are describing; and while use cases offer this too, they are more granular and they describe how the system will react as a result of the actor's action. Both user stories and use cases serve as the start of the collaborative process that takes place in agile projects.

Use cases are not used exclusively in agile environments; they can be commonly found in waterfall projects as supplementary documentation to requirements. In agile projects, use cases can be used to express business, stakeholder, or solution requirements, as well as to support the requirements. Use cases are represented by both graphic and text-based information.

Graphic/Diagram

- Diagramming is a very effective and quick way to represent an overview of solution requirements
- The use case diagram can be read as follows: The <actor> shall have the ability to <perform the feature>
- Each scenario is marked as an oval

Use case diagrams are very effective for situations where there is a need to represent the goals of system-user interactions, for specifying the context and the requirements of a system, and for defining and organizing functional requirements in a system. In general, use case diagrams can help model the basic flow of events.

Scenario

- The text-based portion is expressed as verbs and nouns
- Each scenario in the use case diagram can be expanded textually into a use case scenario

- The scenario consists of the steps that take place in order to complete the scenario
- The use case scenario can be used to represent functional requirements; we need to keep in mind that use cases cannot document non-functional requirements—these typically relate to the overall product and not specifically to a use case and, therefore, we need to document them separately
- The use case scenario can also be represented as a swim lane diagram since the use case scenario demonstrates order, flow, and logic

The components of a use case are listed with explanations in Table 4.5.

Table 4.5 Use case components

Use Case Element	Description
Use Case Number	ID to represent your use case
Application	What system or application this pertains to
Use Case Name	The name of your use case (keep it short)
Use Case Description	Elaborate on the use case
Primary Actor	The main actor that this use case represents
Trigger	What event triggers and initiates this use case
Precondition	What must be true or met before this use case can start
Post Condition	What must be true when the scenario ends
Basic (or Normal) Flow	Also known as the *happy path* scenario, this is the flow that is most likely taken to execute the scenario—it includes the events of the use case when everything is perfect: no errors and no exceptions
Alternative Flows	The most significant alternatives and exceptions; covers other paths or branches off the happy path that can be taken to execute the scenario
Exception Flow	Errors or disruptions that could possibly occur as the scenario is executed

Actor

An actor is a construct that represents the common needs that actual business workers have where the primary actor is the one who receives the value. Many teams with experience in building use case diagrams use this technique to ensure that the full range of actors are represented in the use case—saving the need to go through the process of identifying personas. Figure 4.6 shows a simplified version of a use case scenario and Figure 4.7[8] illustrates a simple scenario that is related to an online shopping system.

Enroll In Seminar

ID:UC17
Preconditions:
  – The student is enrolled in the university.
Post conditions:
  – None

Actor	Major Step	Feature Interaction
Student	Student identifies himself	Verifies eligibility to enroll via *BR129 Determine Eligibility to Enroll.* Indicate available seminars
Student and/or System	Choose seminar	Validate choice via *BR130 Determine Student Eligibility to Enroll in a Seminar.* Validate schedule fit via *BR143 Validate Student Seminar Schedule Seminar Schedule* Calculates fees via *BR180 Calculate Student Fees* and *BR45 Calculate Taxes for Seminar.* Summarize fees Request confirmation
System	Confirm enrollment	Enroll student in seminar Add fees to student bill Provide confirmation of enrollment

Figure 4.6   Use case scenario

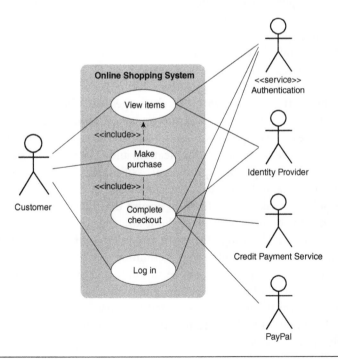

Figure 4.7   Use case diagram

Business Analysis Skills

The BA often leads the effort of putting together use cases and helps the team engage the right SMEs, consider all options, and ensure that no information is missed. Use cases can help define the system boundary by including a subject box in the diagram to define the system scope, therefore, all use cases within the box are *in scope* and entities outside the box are *out of scope*. Business analysis skills (specifically: facilitation, elicitation, and process design) are the key for success in building use cases.

TRANSITION REQUIREMENTS

The final layer of product requirements is transition requirements, which support both the stakeholder and solution requirements. Transition requirements are temporary requirements that take us from the current state to the future state. The transition requirements are temporary in nature because once the project has been implemented, they are no longer needed. Transition requirements include training or data migration. An example of a transition requirement could be: *The credit card application shall be piloted site by site.* It was not ideal to implement the changes across all of the call centers at the same time; if there was a major problem with the implementation, there would be a risk of bringing down all of the call centers nationwide. This is a temporary requirement—once the credit card application is rolled out to all of the call centers, this requirement is no longer necessary.

Project teams typically focus on the solution requirements. This is where the exciting stuff is—along with the challenges and the main focus. However, it is important to ensure that the transition requirements are properly defined and ready. The process of collecting and eliciting information about the transition—the analysis, communication, and validation—is similar to the process for all other types of requirements. With that said, there are still a couple of challenges about transition requirements: (1) they need to be defined after the solution has been defined to ensure that they transition seamlessly from the old to the new solution without any gaps or redundancies and (2) due to the nature and timing of transition requirements, the team is often not focused on these requirements and it is common that there is not enough time to perform the process properly. The results of failing to successfully transition can be devastating and deem the entire endeavor a failure. It is not the job of the BA to perform the work that is related to transition requirements, but the BA must be on guard and ensure that the right resources, capacity, and focus are involved in this process.

VISUALIZING USER STORIES AND AGILE ANALYSIS WORK PRODUCTS

In addition to use cases, there are more techniques to help visualize user stories to help the team, stakeholders, and users better express themselves, and in turn, understand each other. BAs should be familiar with visualization tools and should support the team in introducing these tools—even guiding team members and stakeholders as to how to use these tools. While these are not the only diagrams that BAs should be familiar with, the diagrams discussed in this chapter are specifically useful in the requirements process and to visualize user stories.

Flow Chart

Flow charts are useful for complex and intertwined scenarios. We can use a *simple* flow chart—also known as a process map, process diagram, process flow chart, or a swim lane diagram (see Figure 4.8).[9] A swim lane diagram is a visual depiction of tasks that people perform in their jobs and it describes the steps that people (or systems) take. These diagrams help facilitate conversations during elicitation activities with business stakeholders, identify missing features or requirements during analysis, and track requirements to individual steps within the process flow.

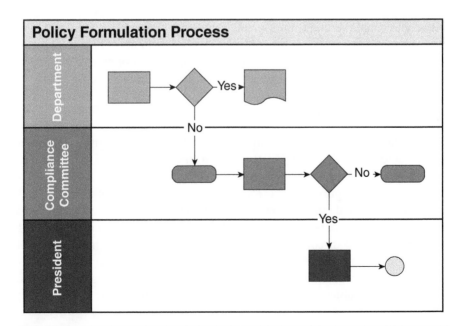

Figure 4.8 Swim lane diagram

Interface Analysis: Storyboarding, Mock-ups, Wireframes, and Prototyping

When flow charts are not illustrative enough for stakeholders and development teams, storyboards provide a visualization of the story scenarios through mock-ups. Storyboarding techniques date back to the 1920s and were used by Walt Disney Studios. Currently, the techniques are commonly applied to systems development and web development. While use cases are the origin and basis for an object model that manages the functionality and business rules of a system, a storyboard focuses on the user—to help the development team better understand the user's needs and develop the right product. Storyboards, like graphical user interfaces (GUIs), communicate design to users with clarity by providing users with real content that is easy to grasp.

Prototyping is about creating a model of the proposed solution. A prototype is also known as a mock-up, which is a representation on a computer screen or user interface that demonstrates how the user will interact with the product or the solution to accomplish tasks or to solve the business problem. Mock-ups are best illustrated visually with wireframes, which can be digital prototypes or as simple as whiteboard sketches. The goal is to keep mock-ups simple since they only serve for communication and will typically be thrown out quickly once the actual product development effort starts. Figure 4.9 shows an example of a mock-up and Figure 4.10 shows a sample prototype for user feedback.

A wireframe (e.g., for a website) is also known as a screen blueprint; it is a visual guide that represents the skeletal framework of a website. The main purpose of creating a wireframe is in order to arrange elements to best accomplish a particular purpose. A wireframe allows a hierarchical definition of design information and it helps make the layout plan according to the specific needs.

Context Diagrams

Context diagrams show the system under consideration as a single high-level process and the relationship that the system has with other external entities (e.g., systems, organizational groups, and external data sources). A context diagram, as seen in Figure 4.11,[10] is also known as a context-level data flow diagram or a level-0 data flow diagram. It is best done on a whiteboard and it helps team members get context and better understand the situation at hand. Context diagrams can help discover user roles and requirements and show high-level interaction of external entities with a subject from a business or system perspective—focusing on data flow. The process to create a context diagram involves: (1) putting the subject at the center; (2) discovering who (i.e., which entity)

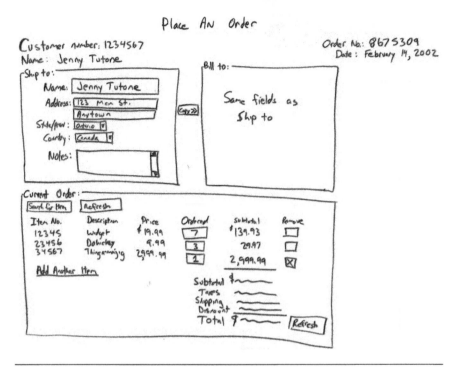

Figure 4.9   GUI mock-up for design

needs to get what information (i.e., reports or data) from the system; and (3) check what information the system needs to provide to the external entities.

Data Flow Diagrams

Data flow diagrams illustrate the relationship between systems, actors, and the movement of data that is exchanged and manipulated over the course of a process. These diagrams help the team relate to requirements through business data objects and processes and are used to help stakeholders and team members understand how data flows through the system as part of the effort to identify data requirements.

Activity Diagrams

Activity diagrams are used during analysis modeling to explore the logic of a usage scenario, system use case, or the flow of logic of a business process. Activity diagrams are similar to flow charts and data flow diagrams.

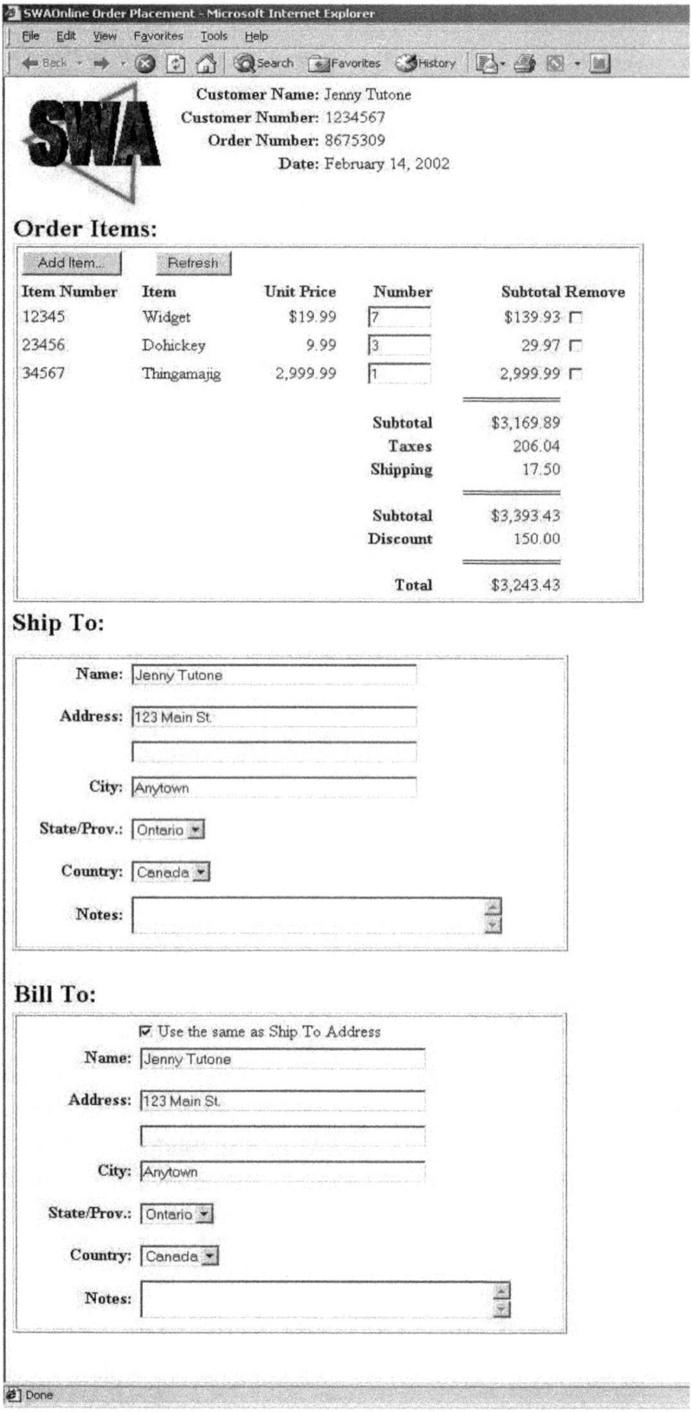

SWAOnline Order Placement - Microsoft Internet Explorer

File   Edit   View   Favorites   Tools   Help

Back ▾ → ▾ ⊗ ⊡ ⌂   Search Favorites History   ⟋ ▾ 🖨 ⊗ ▾ 🖽

SWA

Customer Name: Jenny Tutone
Customer Number: 1234567
Order Number: 8675309
Date: February 14, 2002

Order Items:

Add Item...   Refresh

Item Number	Item	Unit Price	Number	Subtotal	Remove
12345	Widget	$19.99	7	$139.93	☐
23456	Dohickey	9.99	3	29.97	☐
34567	Thingamajig	2,999.99	1	2,999.99	☐

Subtotal	$3,169.89
Taxes	206.04
Shipping	17.50
Subtotal	$3,393.43
Discount	150.00
Total	$3,243.43

Ship To:

Name: Jenny Tutone

Address: 123 Main St.

City: Anytown

State/Prov.: Ontario ▾

Country: Canada ▾

Notes:

Bill To:

☑ Use the same as Ship To Address

Name: Jenny Tutone

Address: 123 Main St.

City: Anytown

State/Prov.: Ontario ▾

Country: Canada ▾

Notes:

🖹 Done

Figure 4.10   Prototype

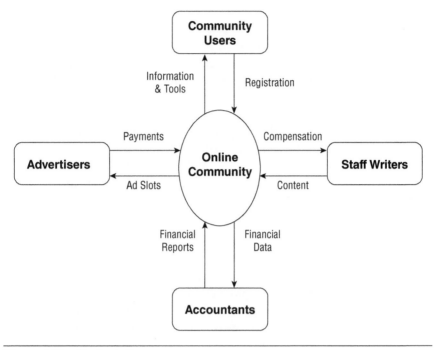

Figure 4.11 Context diagram

Class Diagrams

Class diagrams show the classes of the system, their interrelationships, and the operations and attributes of the classes. During analysis, class diagrams can help represent a conceptual model that depicts the team's detailed understanding of the problem space for the system.

Entity/Relationship Diagrams (Data Diagrams)

Entity relationship diagrams show the main entities, their data attributes, and the relationships between those entities. Like class diagrams, entity relationship diagrams can be used for conceptual modeling and, in many ways, entity relationship diagrams can be viewed as a subset of class diagrams.

Sequence Diagrams

Sequence diagrams are used to model the logic of usage scenarios. Sequence diagrams visually model the flow of logic within a system—helping the team explore and validate the logic.

RECAP

Agile requirements are different than *traditional* requirements—beyond the use of user stories and use cases. Agile requirements typically tend to be *lighter weight* when first identified and they remain ambiguous until more information is required—upon reaching the iteration in which the requirement's features will be worked on and built. At that point, the user story will be fully specified.

A user story is a high-level description of system behavior and it serves as a placeholder for conversation about the requirement. Once delivered, a user story represents a fully functional (even if incomplete) slice of the overall solution.

In *traditional* waterfall projects, BAs define detailed requirements at the beginning of a project, making the assumption that:

- The customer actually knows, and can articulate, what the solution should do at the end of the project;
- Once documented and baselined, the requirements will not change—and if they do, the team will be able to handle the change with minimal impact on the project and product success; and
- The requirements process is confined to a single sponsor who envisioned the product but sits apart from the team. Communication takes place mainly through the BA.

In reality, the customer does not know what they want up front beyond having a problem or an opportunity, a concept, or an idea. Changes to the requirements will trigger project delays, budget overruns, or smaller feature sets and there will be a disconnect between the sponsor and the team. Traditional requirements do not acknowledge the inherent uncertainty in product development that agile methodologies seek to embrace.

This chapter covered the requirements process in agile environments and the contribution of the BA to its success. We also looked at the process of building user stories and at a series of supporting tools that help develop, enhance, clarify, and visualize the requirements—including personas and use cases. Whether the BA skills are performed by team members or by a dedicated BA, they are important for the success of the agile requirements process.

ENDNOTES

1. Project Management Institute *Pulse of the Profession.* (2014). *Requirements Management: A Core Competency for Project and Program Success.* Newton Square, PA: Project Management Institute.
2. http://www.dictionary.com/browse/elicit.

3. Figures and tables in this chapter are adapted from the book *Effective PM and BA Role Collaboration* (2015), J. Ross Publishing, coauthored by Ori Schibi.

4. http://naajaking.com/images/annotated_stories_cases.png.

5. https://tcagley.files.wordpress.com/2013/11/note.png.

6. Gottesdiener, Ellen. Requirements by Collaboration: Workshops for Defining Needs, Addison-Wesley, 2002, ISBN 0-201-78606-0.

7. http://xp123.com/articles/invest-in-good-stories-and-smart-tasks/.

8. https://www.lucidchart.com/pages/uml/use-case-diagram.

9. http://www.jackiewilley.com/home/document-development/process -flows/.

10. http://www.modernanalyst.com/Careers/InterviewQuestions/tabid/128/ ID/1433/What-is-a-Context-Diagram-and-what-are-the-benefits-of -creating-one.aspx.

Web
Added
Value™

This book has free material available for download from the
Web Added Value™ resource center at *www.jrosspub.com*

5

AGILE DOCUMENTATION

Many people, including some agile practitioners, still believe that agile projects do not have or require any documentation. This is one of the most common misconceptions about agile—and if it were up to those people, there would not be a chapter about documentation in this book. Although here too—as in other areas within agile methods—there is no one clear, defined, or formal role for the business analyst (BA) to fill in relation to documentation and reporting. However, there are two functions where the BA can add value within these areas:

1. The BA can help with the documentation and reporting even where team members already own these areas. This is about enhancement of the process, updates, analysis for trends, and ensuring timeliness. The intent is that team members own their pieces of reporting and documentation; but even in these situations, there may be opportunities to refine and streamline the process to achieve more accurate and timely results. After all, with the team being busy producing work products, estimating, communicating with stakeholders, and incorporating feedback into their work, there are many areas where the BA can facilitate and ensure things are done to the right extent. Also, similar to other agile practices, the BA is there to enhance, improve, and refine an already working process—where team members for the most part demonstrate business analysis skills. This is different from waterfall environments where the BA is there to collect items that fall between the cracks due to a lack of business analysis skills among team members.

2. The other area where the BA can add value is due to the tendency of team members to completely overlook reporting and documentation. In a way, we need a BA in the agile environment *because* of the agile environment: the lack of awareness, misconceptions, and a natural tendency to see reporting—especially documentation—as afterthoughts by team members may warrant the need for a BA to pick up some of the slack. With business analysis skills, in general, coming in second to

team members' true *forte*, it is common that team members do not have enough time, capacity, or awareness regarding the areas of reporting and documentation. This is especially pronounced in environments where there is a combination of misconceptions, overworked team members, and teams new to agile. With reporting taking place on an ongoing basis and changes to the way we deal with documentation (specifically— less documentation on a more timely basis), the adjustments teams go through often take too big of a toll on the work performance. This is where the BA can add significant value by identifying the potential deficiencies and complementing the team in these areas.

This chapter covers the areas of reporting and documentation in agile projects and provides context, breaks misconceptions, and introduces areas and ways for the BA to add value.

DOCUMENTATION IN AGILE PROJECTS

Before looking into agile documentation, it is important to review why we even document to begin with. There are valid reasons to document, such as to support communication with stakeholders or with external groups; to address specific stakeholders' needs; or to provide a paper trail for a thinking process, an idea, a plan, or a concept. However, many times we document for the wrong reasons:

- Someone wants to be seen as in control
- Someone wants to appear more important
- Someone wants to justify their existence
- Someone does not know any better
- Someone blindly follows a process that calls for the creation of certain documentation
- Someone looks for reassurances and reiterations that the project is fine
- Someone attempts to specify the work for another group

Agile Documents

What are agile documents and what makes agile documentation different than *traditional* documentation? Agile documents are about maximizing stakeholder investment; they are concise, fulfill a purpose, and describe information that is less likely to change. Agile documents are about describing useful and *good* things to know—they have a specific customer, are designed for the customer's needs, and they facilitate the work efforts of that customer. In addition, agile documents are sufficiently accurate, consistent, detailed, and indexed. There are a few principles to follow when dealing with documentation in agile environments:

1. *Do not document if you do not have to*: challenge the need to create each document that is required by a standard process altogether. One way to ensure the document adds value is by estimating the cost for each document to help the sponsor or product owner decide on the documents' cost-benefit. Regardless of whether we challenge the creation of the document, or the template, the BA can help us in this process, research options, and provide context as to the reasoning behind the need for the document. We must keep in mind that often there will be documents that appear to add no value but are required for compliance reasons. Further, the BA can also support the cost-benefit analysis to ensure that we make the right decision about the document. Finally, we must get permission if we decide to skip standard documents.

2. *For documents that have to be created—challenge the standard template*: although templates are a common and useful way to transfer knowledge, achieve consistency, ensure people do not forget things, and support continuity, templates often contain sections that are not applicable to all projects. Project practitioners often find themselves completing templates that have sections with no purpose and with items that provide no useful information. Here too, the BA can help support the process of challenging sections in templates—and any such challenge must be done with the proper permission.

3. *Document as informally as possible*: project documents typically have two components—content and format. The content is about the information in the document and it is based on findings, research, or updates. The format is about the structure of the document, its style, and its look. One of the keys to ensuring efficiencies and focus in agile documents is to *not* start a document by using an MS Word document or a Visio file. We should start by using the right tool (e.g., Excel or OneNote) to effectively handle information, process content, and formulate thoughts. Before moving forward and creating a document, it is important to think about the consumer of the document, their information needs, the level of familiarity they have with the content, the level of expertise they have with the subject matter, and their position. It is also important to know about the level of support they have for the initiative and the level of influence they have on the matter at stake. We should also consider legal or audit requirements for using standard templates, as well as determine how informally we can document while still meeting the consumer's basic needs and constraints. The BA can

save the team time by providing legal and compliance information and insights about the needs of the document's consumer.

4. *Procrastinate*: agile accepts that there is ambiguity in projects (especially early on in projects) and it welcomes ambiguity until further information is needed. Due to ambiguity, we need to be able to make decisions as late as responsibly possible based on the timing of the information we receive. We can continue the same line of thinking with documentation—document as late as responsibly possible in order to avoid errors due to insufficient information or misunderstanding; reduce the need to maintain documents and keep them current; and minimize the creation of documents that are not really necessary. It is important to keep in mind that procrastination does not mean delaying documentation just because we have a habit of delaying things and it is also not about leaving things until the last minute. Those of us who are procrastinators should not celebrate this habit and think that by being procrastinators they are agile—you must know the difference.

There are many forms of agile documents including (but not limited to) use cases, graphical user interface (GUI) models for business requirements, GUI mock-ups for design, and prototypes for user feedback. Even when team members have business analysis skills, quite often they do not have modeling and diagramming techniques that will allow the BA to support them with the creation of or updates to the models, diagrams, and documents. The majority of agile documents are not necessarily used exclusively by agile projects and can be applied in any type of environment.

Agile documents can be broken down into four groups based on the planning cycle and where we are in the project: project backlog, iteration backlog, list of impediments, and *definition of done*. Table 5.1 shows the breakdown of the documents by each of these categories. The BA supports the documentation process through coordination with team members and the product owner, covering areas that need additional support (e.g., reprioritization, estimating), capturing feedback, and coordinating items that are *parked* in the shadow or defect backlogs.

Agile Documentation Process

The documentation process in agile projects is iterative and essentially repeats itself based on the project cycle. It is done in the background and it supports the product development process:

1. It starts with a list of tasks that are captured in the backlog
2. A team member (e.g., developer) updates the content of each story they work on

Table 5.1   Agile documents: four groups of documents based on the planning cycle. The BA supports the reprioritization process, helps with estimating if needed, ensures velocity is watched, and keeps an eye on the *secondary* (shadow and defect) backlogs

Project Stage	Document Type
Project backlog	• Prioritized and indexed list of user stories (ongoing) • This is the *what* of the product/solution • Open for collaboration with decisions by the product owner
Iteration backlog	• A list of tasks for the user stories identified and prioritized into the iteration • This is *how* the work is going to be performed • The list is estimated and planned for by the team and is owned by the team
List of impediment	• A list of issues and impediments (team/product/ organizational) — with proposed resolutions
Daily task	• A list of things to do for each team member — managed by individual team members — based on the individual's work
Definition of done	• A checklist of acceptance criteria for the product increments

3. Another team member (e.g., another developer or the BA) provides a peer review of the technical content for accuracy and completeness

4. A technical writer (e.g., could be the BA) performs quality assurance (including glossary and content)

5. Someone with a role equivalent to the delivery manager accepts changes and prepares for delivery

6. The final step is announcing to the customer that an item is ready for delivery

The documentation process in agile projects is different than the one in waterfall or traditional environments. While in traditional methods we spend significant effort on detailed documentation early on, followed by a significant effort to keep these documents current, in agile methods we end up spending less effort overall on documentation. This is due to the high-level documentation that takes place early on, support of deliverable documentation throughout, and system overview documentation at the end. Whether the agile team delivers documentation continuously or late in the project, the total effort spent on documentation in agile projects ends up being significantly less than it is in traditional methods. Figure 5.1[1] illustrates the differences.

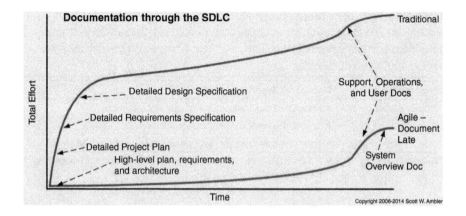

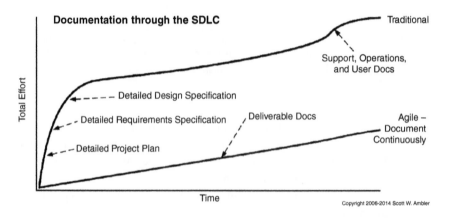

Figure 5.1   Documentation through the software development life cycle (SDLC): whether agile methods deliver documentation continuously or late—we end up spending less effort on documentation in agile projects than we do in traditional method projects

Too Much, Too Little, and Compliance

With agile focusing on value creation and specifically with documentation—minimalism and less formality—there is a fine balance to maintain between too much and too little documentation. That right balance is far off from having no documentation altogether, but we must resist our natural tendencies (and the natural tendencies of the stakeholders) to over-document (almost as if we document *just in case* we will need it later). Another common challenge with agile documentation is when we operate in high compliance environments. These environments require more formality in documentation, but even then we should document only what we must. Reduce document size as much as possible and even challenge the need to complete unnecessary sections—but all of this has to be done while ensuring that we do not break the rules. Regardless of the level of compliance allowed, we should always seek clear and formal approval for skipping any documents or modifying standard templates. The BA is often familiar with compliance, audit, and other rules around documentation, and he or she can provide guidance in these areas.

Distributed Teams

When teams are not colocated, it naturally increases the need for documentation. Without the use of face-to-face meetings, distributed teams require more formality and with that comes more documentation. The added documentation will not completely reverse the slowdowns and limitations due to the distance, but it can help—especially with support from technology that helps make things more interactive.

Use technology to simulate face-to-face meetings as much as possible. Commonly used technologies can be utilized to help facilitate the challenges introduced by distance:

- Instant messaging
- Screen sharing
- Webinars
- Virtual white boards
- Teleconference calls
- Video conferencing
- Online document management system, sharing, and repository

Level of Formality

It is important to identify clear criteria for the team regarding the level of document formality so the team refers to and uses different documents consistently and in the right context. For text-based documents, instant messages and office messenger types of communication are the least formal; e-mails and memos are higher in formality but still, for the most part, are considered as an informal style of documentation; and traditional documents and templates are the most formal of the text-based type of documents. As for graphical-based documents, whiteboards are the least formal, followed by memos with sketches, and going up to Visio and other formal diagrams. While these rankings are not an exact science, it is important that the team shares a consistent view of the associated or implied level of formality for each type of document. When in doubt, team members should seek support from the BA—including support in drawing or using certain types of documents.

AGILE REPORTING: ARTIFACTS, PROJECT REPORTS, AND INFORMATION RADIATORS

Agile reporting, in a similar fashion to all other areas within agile methods, focuses on efficiencies, but at the same time, comes to ensure that all stakeholders get the information that they need and that metrics are in place for processes and project deliverables. Agile reporting is supported by metrics that optimize the delivery of value to the customer, help understand the development process, and makes the release of the product easier. The goal is to produce information *radiators* that *immerse* the team and the stakeholders with information that is accurate, timely, and above all—relevant. The following sections will describe the commonly used metrics in support of agile reporting.

Iteration Burndown

Agile teams work within time-boxes. These are consistent and defined periods of time for an iteration's length (commonly the iteration length is 2–4 weeks, but it is not mandatory to follow these time-boxes). The iteration burndown report tracks the completion of work throughout the iteration against the estimates the team provided at the start of the iteration. The x-axis represents time, and the y-axis shows the amount of work left to complete, measured by story points or hours. The goal is to keep the actual performance in line with the plan since it serves as an indication that the work the team is producing is keeping up with the forecasted work. Clearly, the goal is to have all of the work that was planned for the iteration completed by the end of the iteration. The BA can help ensure that team members continuously report on their progress (or impediments) and that the information reported is not *fudged*. The intent is not for the BA to be a babysitter for team members who do not do what they need to do (including reporting), but with team members focusing closely on their work, they may forget to report on progress or to update task boards or charts in a timely manner. If the reporting does not reflect actual progress, it may distort the information about the team's velocity, hurt learning, compromise improvement opportunities, and ultimately reduce the amount of value produced.

It is important to look for common *anti-patterns* such as:

- When the team repeatedly finishes work early because it does not commit to enough work
- When the team repeatedly fails to finish the forecasted work because it overcommits to too much work
- When work changes midway through the iteration—these changes come from the product owner and include changes or additions to the iteration's scope

- When the burndown chart fluctuates too much because work has not been broken down properly

Velocity

Velocity is the average amount of work that the team completes during an iteration. It is measured by story points or hours. Velocity is not only the main indicator of the team's performance, but it is also useful for forecasting. Within 1–3 iterations into a project, teams usually get into the *rhythm* and start producing work in a consistent manner. With consistency, it becomes easier for the team to forecast its performance moving forward. The consistency can be achieved if it is the same team moving forward for a similar type of work; changes in the team makeup will cause fluctuations in the velocity. The velocity also helps the product owner predict how quickly the team can work through the backlog because the report tracks the forecasted and completed work over multiple iterations. For example, if the product owner wants to complete 700 story points in the backlog, and the team generally completes 70 story points in one iteration, the product owner can figure out that the team will need around 10 iterations to complete the required work.

Velocity can change over time even within the project. It is normal to expect an increase in velocity as the team optimizes relationships and processes, but only to a degree. There should also be considerations around a certain calendar period for reduced velocity, as well as for fluctuations due to organizational events, resource allocation, and performance. The latter includes slowdowns due to the need for bug fixes (e.g., we should expect a lower velocity in the first iteration after a release) or due to testing activities, especially when there is no sufficient test automation. Otherwise, with no external factors affecting the team, a decrease in average velocity is usually a sign that some part of the team's development process has become inefficient and should be brought up at the next retrospective. The BA can provide insight into any conditions or events that may impact the team's velocity.

The team should also watch for significant fluctuations in velocity, which may indicate some issues around the team's estimating practices.

Epic and Release Burndown Charts

Epic and release burndown charts track the progress of the work over a larger body of work—an epic or a release. When teams have several epics, it helps to track both the progress of individual sprints, as well as that of epics within the release. The units of measure on the y-axis are commonly expressed by story points. Similar to traditional methods, the team and the BA need to be on guard for *scope creep*, which is the injection of additional requirements into the previously defined work. While it is common that new needs are introduced throughout the project, they have to be introduced in an organized fashion (and prioritized into the backlog). With that said, we need to keep in mind that agile projects are about adapting to change since the change represents new and additional value for the customer. The release burndown charts help keep everyone aware of the progress toward delivering the product increments of the release. The team and the BA need to be on guard for chronic scope creep when progress is not made for an extended period and when there are no updates on the release burndown chart.

Burndown charts all follow the same idea: there is a forecast line that trends from the upper left side of the chart to the lower right, representing the amount of work that needs to take place at any given time frame based on the projected velocity. The second line represents the team's actual progress. There can be a project burndown chart (showing typically by story points, the work remaining or iterations); a release/epic burndown (by iteration); and an iteration burndown chart (typically showing by workdays, work hours, or story points). Figure 5.2[2] provides an example of these charts. The team updates the progress on the chart, but the BA can help ensure that progress that is reported is actually made.

The vertical axis is the amount of work and is measured in units that are customized to the individual project. Some common units are the number of tasks and estimated hours or story points (in agile project management methodologies). The horizontal axis is time—usually measured in days.

Similar to the use of burndown charts, some agile teams use burnup charts to track progress toward a project's completion. In its simplest form, a burnup chart contains two lines: a total work line for the project scope to complete and a work completed line. The burnup chart shows both completed work and project scope; and the project will be completed when the lines meet. Unlike the burndown chart that shows remaining work and the downward trend toward the goal of a remaining zero balance of work; in the burnup chart, as shown in Figure 5.3,[3] the count starts at zero and the lines work their way up until they meet at the completion of all work.

Cumulative Flow Diagram (CFD)

The CFD helps ensure that the flow of work across the team is consistent. With the number of issues marked on the y-axis and time on the x-axis, different colors indicate the various workflow states. This type of diagram provides a visual indication of bottlenecks and shortages in conjunction with *work in progress* limits. Ideally, the CFD should look fairly smooth (moving from the left to the right). Bubbles or gaps in any one color indicate shortages and bottlenecks that need to be addressed. Common anti-patterns to look for include blocking issues

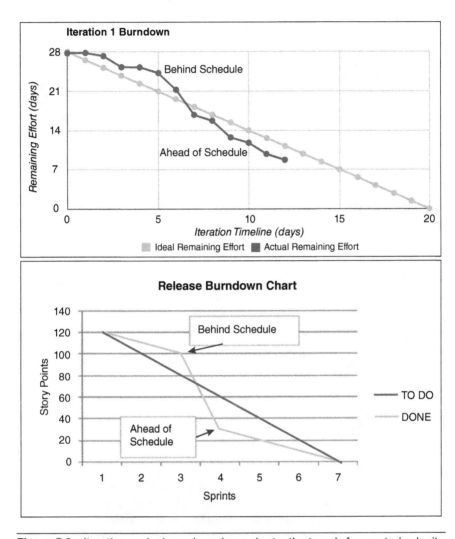

Figure 5.2   Iteration and release burndown charts: the team's forecasted velocity determines the slope of the projected line and the team's performance is reflected by the actual trend line. The timelines for the project or release burndown charts show progress by iteration; and the timelines for the iteration burndown charts are typically by day within the iteration.

that create a backup in parts of the process and starvation in other parts; and hanging issues—where the product owner does not close issues over time. This unchecked backlog growth may be as a result of low priority, or even obsolete issues, but since they are not closed they distort the information on the diagram. Figure 5.4[4] illustrates the uses and benefits of the CFD.

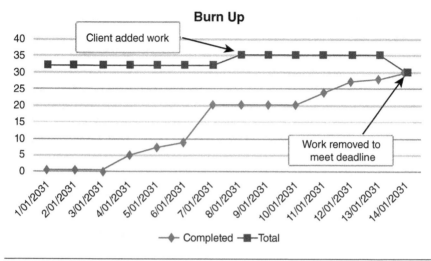

Figure 5.3   Burnup chart: the burnup chart has one advantage over the burndown chart—it shows the scope line and tracks when work is added to or being removed from the project

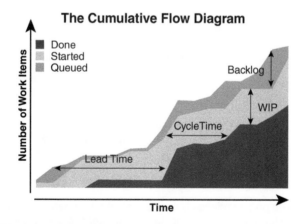

Figure 5.4   Cumulative flow diagram: a stacked area chart showing (by time intervals) the number of items in each stage of the process. As time goes by, the chart shows the flow of items through the process in a cumulative fashion and not by incremental change.

Control Chart

Control charts illustrate the cycle time of individual issues—showing the total time it takes items to make it from *in progress* to *done*. Teams with shorter cycle times tend to produce more value and teams with consistent cycle times across many issues are more predictable in delivering work. Measuring cycle time can help improve the team's processes with the goal to have a consistent and short cycle time—whether the work involves a new feature, technical debt, or anything else. What the team needs to look for in control charts is trends, but there is a need to keep an eye out for increasing cycle times when there is no expansion in the definition of done and erratic cycle times.

Task Board or Kanban Chart

The information about the iteration's work and progress is visible by putting it on a task board or a Kanban chart. Team members update the task board continuously throughout the iteration based on the story that each team member works on, the tasks that are in progress, and subsequently, the status of these tasks. Figure 5.5[5] shows a sample of such a board. All the user stories and tasks that are identified for the iteration appear in the *To Do* column of the task board and as team members make progress, they move the respective tasks on the board from one column to another based on the status of the work they are performing. Different teams identify different names for their columns, but typically, from left to right, the columns will include the user story (Story column); the associated tasks to perform in order to complete it (To Do column); and an *in-process* column (In-Process column). It is common to also find a column that is named *hurdles* or *impediments* for any tasks that face issues that need to be handled. As we move right, we will find a column that represents tasks and

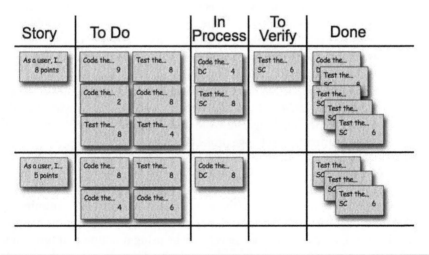

Figure 5.5   Task board or Kanban chart: Each row on the task board is a user story. The team identifies all stories to be performed in the iteration and these stories and tasks are each represented by a sticky note in the *To Do* column.

stories that are ready for testing (To Verify column), and to the right of it there will be a column named *Done* or *Complete*. Team members move cards on the board (which can also be electronic boards) either during or before the daily stand-up meeting where estimates are adjusted.

More Reporting

One of the main challenges with the way we report in traditional method projects is that the reporting is open for interpretation and when we talk about % complete, it can mean any one of the following:

- % estimated work effort complete
- % work products complete
- % tasks complete
- % customer-requested functionality/content delivered
- % budget used
- % schedule used
- % that I think you will accept as reasonable

In addition, a weakness in the metrics used in traditional methods—including earned value management (EVM)—when encountering high-change situations is that the actual progress may be relatively reduced every time there is a change to the project scope. Also, it is easy to be plagued by the 90% syndrome where the team reports a 90% completion of the project during the testing phase even in cases where it takes significantly longer to complete the project. The use of most reporting tools, including MS Project, typically cannot adapt to this reality of changing conditions and growing scope—even if these changes and additions are controlled and recorded properly. Although EVM is very good at building trust, measuring the budget and schedule risk level, and communicating progress, it also has some weaknesses. It is possible to make progress first on easy tasks while ignoring important tasks—but these metrics will not show any indication of these practices. In addition, EVM metrics do not warn anyone that the team is building the wrong thing, and EVM loses most of its benefits when your baseline keeps shifting (high change projects).

Back to Velocity

In addition to measuring velocity (the team's rate of progress) by summing up the number of story points or ideal days that the team completed in its most recent iteration, we can also measure velocity by using function points or some other metric that includes a measure of complexity. There is still another way to measure velocity—as a percentage (number of story points completed, divided by the number of story points planned). This provides the team with a metric for estimating accuracy and efficiency. Velocity percentage can be used to adjust the forecast schedule across multiple future iterations, and overall, velocity is used in a similar fashion to EVM's schedule performance index metric. Velocity is part of agile earned value.

RECAP

Project documentation and reporting are critical to project success in any methodology, including agile. When working in agile projects, it does not mean there is a free pass concerning documentation. Beyond legal and compliance reasons, there can be no proper planning or estimating process without documentation. Further, there can be no sufficient tracking, controls, and reporting. With efficiencies in mind, agile methods call for reducing documentation to the minimum necessary; doing it informally, if possible; and delaying the creation of documents to the latest possible responsible time. The BA supports the team in creating and maintaining project documentation, but the BA is not there to replace the need for team members to do their role as part of documentation.

Reporting is important because it provides updates about the agile project, but that information needs to be accurate and timely. Information radiator is a generic term for any type of reporting document that the team places in a visible location (physical or electronic) so that all team members and relevant stakeholders can see the latest information at a glance. The information radiators contain a variety of artifacts and charts that are visual, easy to understand, intuitive, and indicative of the actual progress—and are updated promptly by the team.

Agile methods do not rely on the BA to perform the documentation or the reporting for the team, but since these two elements are done differently in agile projects, the BA can provide support in updating, ensuring timely and accurate information is in place, and providing references and help for team members when dealing with documentation and reporting. In addition, the BA can provide expertise about the need to challenge certain documents or their components, and the BA can also help create models and diagramming techniques that support documentation or reporting in the project.

Finally, agile documents and reporting also focus on simplicity, and depending on where we are in the project, there are *cyclical* elements that take place to maximize the value created for the customer. These cycles not only provide context on the extent and type of documentation and reporting required, but they also serve as checks and balances pertaining to: the project backlog, iteration goals and theme, iteration backlog, and the daily tasks of team members. Documentation supports reporting as part of agile ceremonies, including:

iteration planning, daily stand-up meetings, the iteration review, and the iteration retrospective.

ENDNOTES

1. Adopted from http://agilemodeling.com/images/lifecycleDocumenta tionContinuous.jpg.
2. https://4squareviews.files.wordpress.com/2016/02/burn-down-chart. png for the release burndown chart and http://www.softwaretestingstu-dio.com/wp-content/uploads/2017/03/Release-BurnDown-Chart.png for the iteration burndown chart.
3. http://www.clariostechnology.com/productivity/blog/whatisaburnup chart.
4. https://i.stack.imgur.com/ebDym.png.
5. https://www.mountaingoatsoftware.com/uploads/articles/MockedTask Board.jpg.

6

THE BA ROLE IN PLANNING
AND ESTIMATING

The way agile methodologies perform planning and estimating activities is different than the way they take place in waterfall environments. Beyond the fact that the actual acts of planning and estimating serve as team-building exercises, the effort is inclusive and collaborative. Further, these activities yield more consistent and realistic estimates that, with time, are proven to be more accurate. The planning process is done in such a way that it allows for multiple checks and balances since different *rings*, or *layers*, of planning take place at different times—including product vision, road map, and release; iteration; and daily plans. The planning process becomes more specific and detailed as we get closer to the time of performing the work we planned and each layer validates the previous set of plans. Agile planning and estimating truly bring to action the concept of rolling wave planning where high-level plans are elaborated and validated—one iteration at a time. The effort around planning and estimating involves stakeholders, the product owner, and all team members; and it is crucial that all team members have the relevant skills to perform these activities. Agile projects can benefit from having a business analyst (BA) to support the planning and estimating process and to complement team members' planning and estimating skills. This chapter deals with the planning and estimating processes—discovering ways to refine them and looking at areas and timing where a BA can help produce more realistic and consistent estimates and plans.

A NEW ROLE FOR THE BA

Before getting to the *tactical* elements of planning and estimating, let's take a look at another area where the BA can add value to agile projects. This is through supporting the product owner in more a strategic manner than just bridging between the product owner and the team. Although for many having the BA supporting the product owner may be an afterthought, other practitioners view the BA's work with the product owner as a natural fit, and some even believe that the BA can outright replace the product owner.

1. *BA supports the product owner:* This is where the BA bridges, clarifies, and provides context for the product owner, for the team, and for the two-way communication between the product owner and the team. In such a situation, the BA helps consolidate questions, provides clarifications, performs research, finds expertise, presents options, and ensures that the product owner and the team communicate effectively and efficiently with each other. The BA provides context and ensures that all options are taken into account in the prioritization process. The BA also ensures that actual performance, changing needs, knowledge obtained, and constraints are all considered when reprioritizing and taking action. The BA is a product expert, a business expert, a user-needs expert, and a communication enabler. When the product owner is not sufficiently involved, the BA can play a key role in ensuring that knowledge is still shared and actions are taken based on the latest information. The contribution of the BA in this setup can be significant, but is limited to the confines of the users' needs (as identified by the users).

2. *BA replaces the product owner:* This is taking the role of the BA one step further—when the product owner is not available and cannot provide the necessary feedback, some practitioners see the BA as a natural replacement. While it may look tempting since it takes the role described in the previous paragraph further, there is a clear consensus that it rarely adds value. The BA does not have the authority, seniority, clout, experience, or context to outright replace the product owner and this kind of arrangement often ends up failing. One way or another, the benefits that a BA can introduce to the project in these two aforementioned roles (especially in item #1 as a support for the product owner) are limited, but the potential benefits that a BA can add to projects in the following setup is more significant and with the potential for a much higher impact.

3. *BA enhances the focus of the product owner:* An increasing number of practitioners and business stakeholders question, in an interesting and yet surprising trend, whether agile projects actually produce more business value for the customer. The answer to this question seems to be— no, agile does not increase the total amount of business value it produces for the customer significantly beyond that of a waterfall. However, the evidence about other benefits that agile projects produce is compelling:

   • Embrace change and enhanced ability to adapt to changing needs and conditions

- Deliver value even when the end goals are unclear
- Deliver value earlier and faster overall than waterfall
- A better chance to deliver quality
- More checks and balances and opportunities for customer feedback throughout the project
- Strong interactions, clear communication, and more collaboration than in waterfall
- The voice of the customer is heard loud and clear
- A state of mind of continuous improvement

Alongside the benefits, there are some down sides:

- Less than concrete plans
- Ambiguity (until more information is needed; nevertheless, stakeholders do not like ambiguity)
- There is a need for specific team skills
- Less documentation may increase risk
- More commitment is required by team members (and testers)

With that said, there is no question that agile helps deliver project success when it is applied in the right way and within the right context.

Despite all the improvements agile projects deliver, the chatter about a lack of increased value creation is not about questioning the benefits of agile or challenging the premises on which it is built. There is no question that agile projects are better at delivering the intended value to the customer and that agile projects fail at lower rates than waterfall, as illustrated in Figure 6.1; however, the

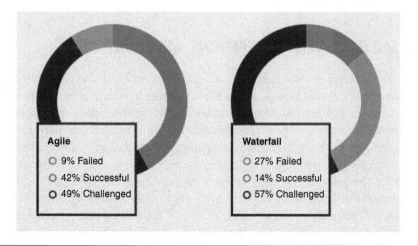

Figure 6.1   Agile delivers value

question that is introduced is whether agile helps increase the amount of value we produce for the customer *beyond* the potential that has been identified. Are we entering a new territory of value that was not identified or articulated before? The answer to this question is no.

Agile: Not Unleashing New Potential

When looking at the potential reasons as to why agile does not take us to the next level in creating value—the level beyond what the customer asked for—one cannot help but notice that while agile projects improve on the service delivery, the *heavy focus on users' needs* means that the team and the project are confined to those needs that are identified by the users. As a result, the project team often ends up missing out on opportunities that go beyond what the user identifies and states. This means that agile projects are *stuck* and constrained by user paradigms without the ability to see beyond what the user defines and states. It is similar to the concept of: *I don't know what I don't know, and I don't know that I don't know it because I don't know what I don't know.* In more simple terms, agile continues the trend from waterfall, which is failing on delivering more meaningful business value. While it is done more collaboratively with early realization of value and more responsiveness to change, the bottom line is still the same—falling short of its potential to deliver value.

It is clear that agile manages to produce a larger and more focused portion of the value identified, but the challenge is about the value that has not been identified. Unleashing the ability to identify and articulate areas of value that were previously unknown or unclear can significantly improve the ability of projects to deliver more value for customers.

SAME BA SKILLS—DIFFERENT APPLICATION

The idea behind the concepts of enhancing the focus of the product owner comes in addition to the need of the BA to support the team. However, *enhancing* the focus of the product owner is truly about adding value as opposed to many of the *tactical* support activities that the BA is expected to perform that are intended to overcome skills shortcomings and address coordination and touchpoint issues within and around the team. The way the BA enhances the focus of the product owner is through application of the BA's analytical and research skills. These skills will help reach out to users and engage them in such a way that not only ensures the creation and delivery of value for the users' needs, but also gets the users to open up beyond their box to look for new needs and opportunities for value creation. For example, when a manufacturer asks for an app to help identify cases of counterfeiting, the performing organization should challenge the scope of the needs to check whether the same set of data and capabilities can also help the manufacturer collect meaningful information about its customers. In short, the BA does not replace the product owner, but rather leverages the knowledge that all sides have and explores options for the creation of new and additional benefits beyond those that have been stated. This is not about scope creep, or gold-plating—as the team does not produce anything for the client in addition to the scope of work that was defined—however, this is a process that enables the opportunity for an increase in overall value should the client decide that it is in scope.

With more potential benefits and value identified, the door is open for more collaboration, more opportunities, and more innovation, which in turn, will serve both parties—the user/customer and the performing organization.

Application of Business Analysis Skills

The business analysis skills that are applicable for this enhanced role toward value creation are not any different than what we look for in BAs in both traditional and agile environments.

- *Problem solving*: Understand a problem from all perspectives, analyze and investigate available options and constraints, and identify options for a solution. This allows for a better understanding of the underlying issues that impact the user and provides the user with options not only about fixing the problem, but also looks for additional related areas of potential benefits.
- *Communication*: Be expressive and help articulate the mandate, user needs, options, and constraints. Being an effective communicator goes well beyond getting one's own points across; it is about being able to understand what is being said (listening) and what is not being said (unidentified needs, unrealized options) as part of analytical skills.
- *Management skills*: This is not about overstepping boundaries—those of the product owner or the Scrum Master/project manager (PM)—it includes areas related to the management of stakeholder engagement, requirements, identification of needs, deliverables, deadlines, and context.
- *Research and investigative skills*: Many projects have solutions and functionalities that are predefined without thinking of whether they are all applicable or whether there are additional needs. The preconceived notions we have *lock* us within a paradigm and create tunnel vision that prevents us from seeing beyond the stated solution. These skills help us

start by identifying the true problem and recommending a solution that requires analysis, which warrants investigating and research skills.

- *Technical skills*: While the product owner typically has product information, the BA can complement this with relevant technical skills. With the product owners' business knowledge and context, the BA will be able to become more of a conduit between all stakeholders involved—especially between the business and the technical team.
- *Breaking the paradigm, pushing the envelope, challenging the status-quo*: The product owner represents the need and the mandate for change, but the product owner often lacks specific knowledge of processes and even constraints. The BA, with more visibility of processes, constraints, and elicitation techniques, can be the engine behind ideas for change by introducing options, potential ideas, and improvement areas that go beyond the stated needs. At the same time, the BA can help the product owner, users, and the team remain objective in the sense of understanding constraints, limitations, and dependencies.
- *Decision making*: The BA can provide context and time sensitivity, including access to benefit-cost analysis, to help make informed, timely, and correct decisions. The BA can own some of these decision areas in relation to certain levels of change, preventing scope creep, or determining an option's viability.
- *Creativity and innovation*: It is important for benefit realization planning, solution assessment, and process improvements to be creative and to see beyond the obvious so we can present options to the users and the decision makers for consideration.
- *Enterprise analysis*: The BA is one of the common threads from when an opportunity is considered and on through the life cycle of a project, and as such, can provide context and opportunities that include potential process improvement for the customer as part of the solution that is offered.
- *Efficiencies and eyes for waste*: These skills are applicable for any part of the project, but in this context, it is about introducing potential efficiencies for the customer and the user and to reduce waste.
- *Asking intriguing questions*: A lot has been said about what makes a good question and what types of questions should be asked at different stages and circumstances. However, the BA is the natural candidate to ask intriguing questions—these are questions that allow ambition, innovation, and outside-of-the-box thinking to meet constraints and reality. These questions get people to think, introduce new options, and consider the merits of situations. Even if there are team members who have business analysis skills, these are usually focused on the day-to-day activities of the team (within the solution). Further, the capacity of the team

members with those skills is often limited and is rarely, if at all, utilized to enhance the product owner's focus.

BUSINESS ANALYSIS ACTIVITIES TO ENHANCE THE FOCUS OF THE PRODUCT OWNER

Later in this chapter, we cover the contribution of the BA to enhancing the team's focus and to improving their efficiency and productivity through effective planning. In fact, the majority of this book's focus is on the *tactical* areas where the BA supports the team. However, the BA's role in enhancing the product owner's focus is more strategic and of a higher potential impact than the *traditional* BA role in agile projects. The BA's value creation as part of the agile planning cycles is important, yet it is confined to the value that has been identified. Further, the introduction of a BA in order to enhance the product owner's focus, as discussed in this section, virtually cancels the arguments *against* the involvement of a BA in agile projects. Recall that we should not confuse these concepts with those who call for the BA to outright replace a product owner or with those who call for the BA to *cover* for an absentee product owner. While these two areas (replace and cover) are helped when the BA bridges, asks questions, and provides clarifications to and from the product owner, this is not the core intent of the BA's involvement in agile projects.

The approach discussed in this section opens the door for the BA to add value for the customer on a strategic level (looking, researching, and investigating the potential for new value), instead of having the BA contribute to the project at the level of supporting team members by holding their hands on a day-to-day basis. The strategic value creation approach that is part of enhancing the product owner's focus can span one or more of the following areas:

1. Looking for needs and opportunities that the user has not realized.
2. Breaking the paradigm that what users say is what they want and recognizing that the user may miss certain points of view. This consideration goes well beyond the application of elicitation techniques because elicitation techniques are about ensuring that users understand and articulate their actual needs—here we try to look for unrealized needs.
3. Helping to define and articulate what constitutes business value and communicate it across the project organization.
4. Defining how to measure business value and benefits.
5. Focusing on understanding the drivers, values, and goals of the users before digging into the features, requirements, and solutions.
6. Looking beyond the user to discover if there are any needs that other stakeholders have that we are failing to address and to consider potential

non-users who do not use our product because their needs are not addressed. A generic product development process involves three stages: research, discovery, and product. Most projects naturally focus on the product with some attention to the discovery process. However, the BA can help the product owner not only shift more focus toward the discovery stage, but also to research. This will expand the range of options available.

7. Demonstrating a cross-functional collaboration and organizational agility mindset in order to improve the organizational ability to leverage from benefits across departments and to streamline the creation of value in order to remove silos and organizational boundaries. This can also help the product owner identify potential areas of resistance and proactively address the needs or make adjustments to understand the source of the resistance and reduce it.

The business analysis activities discussed here have the potential to produce a wide range of benefits, including the potential to increase value-creation for the customer. This opens the door for more opportunities for the performing organization and maintains long-term thinking. Long-term thinking is not only about prioritization, it also reduces the amount of waste due to short-term shifts in priorities and attempts to realize short-term savings.

Now we can move to the more obvious and tactical roles of the BA in the agile planning and estimating process.

TRADITIONAL PLANNING

The planning process in waterfall, also known as the *traditional* method, is extensive and long; however, it does not typically yield plans that are realistic or reliable. One of the main challenges with traditional planning is that it takes place early on in the project, attempting to know what is going to happen later on. With long-term planning, the team invests significant effort and cost into planning activities for things that may change, only to subsequently spend significant effort in changing those plans later. In addition, traditional methods focus on the completion of tasks rather than delivering features, despite that customers get little value from the completion of tasks since features are the true measure of customer value.

In addition, when we review traditional task-based schedules, we tend to look for forgotten tasks rather than for missing features; and with little effort at feature prioritization across multiple releases, we tend to build all features in one release regardless of the importance or value that these features produce to the business. Figure 6.2 illustrates the high number (64%) of features that are built for no reason or are rarely used by the customer.

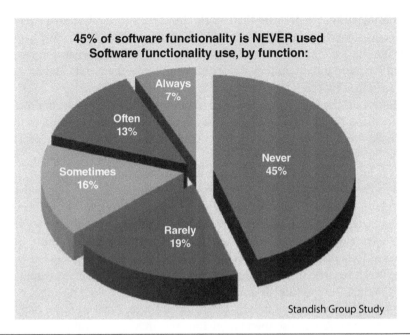

45% of software functionality is NEVER used
Software functionality use, by function:

Always 7%
Often 13%
Sometimes 16%
Never 45%
Rarely 19%

Standish Group Study

Figure 6.2   The cost of traditional planning: feature usage within deployed applications shows that close to 64% of all features are never or rarely going to be used

Another flaw of traditional planning is the fact that we end up approximating our estimates despite the attempts to provide confidence about their accuracy. The approximation takes place when we do not really know, in detail, what is going to happen with each individual task that is planned so we rely on the law of averages and hope. We can break down tasks into three general types: (1) tasks that finish early; (2) tasks that finish on time; and (3) tasks that finish late. At that point, many team members and stakeholders apply the law of averages and believe that some tasks will finish early, other tasks will finish on time, and some will finish late; however, it will all average out at the end. This is a common and damaging mistake. Even if we are lucky and a task finishes early—due to the resource sharing reality of most organizations—we will not be able to benefit from it or to make any gains since the next resource who is assigned to work on the next task is not going to report to the project until the original time they were supposed to. The result is that the project (or at least the path of the activity that finished early) will now stall and wait for the resource to report to the project, and any gain that was made from finishing the previous task early is canceled.

Further, it is important to come to terms with the reality that tasks rarely finish early. In fact, if team members inflate (or pad) their estimates to ensure they get sufficient time to perform their work, the effect ends up backfiring. Parkinson's Law states that *work expands to fill the time available for its completion*, or that *the fish will grow to the size of the tank*. It means that when the resource asked for more time than they thought it would take them, it is human nature and therefore expected that the resource will now work more slowly so that the work fits into the time frame allocated to perform it. For example, if a resource knows that a task should take them five days to perform, but they ask (and get) seven days to do it, they are not only going to fail to benefit from that extra time, but in fact it will backfire. People's natural tendency is to not work as fast when there is lots of time available for a task and an official schedule showing seven days to complete a task gives them implicit *permission* to take that full week. Therefore, it is safe to say that the chances that the resource will deliver their product late after the seven-day period are higher than the chances they would have been late after five days. While this might sound odd, when more time is allocated for a task than is needed, there is a greater chance of being late than if there was a realistic (but more aggressive) and shorter estimate to complete it. In addition, if the resource is done performing the task early, he or she may attempt to *gold-plate* the solution by adding extra features that were not requested by the sponsor. People also spend more time researching novel approaches to the solution if they perceive that they have lots of time. Finally, the resource may also be concerned that if they finish a task early, they may be accused of *padding* their estimates or that they will be expected to complete their other work early as well. Any way we look at it—inflating an estimate does not add value to the project, and even if there is a slight chance that a task finishes early, the team and the project will not end up benefiting from it.

Delays Are Nonlinear

There are more issues with traditional planning because they are task-based plans. These plans focus on dependencies between tasks and late completion of one task can cause subsequent tasks to also be late. Since delays are nonlinear, a one-day delay in completing a task may result in a longer delay downstream (e.g., due to missing a testing environment or missing a resource availability window). Considering the previous point that tasks rarely finish early and the reality that the delays are not linear in traditionally planned projects, testing rarely gets to start early, and a timely completion of the testing effort is rare.

Multitasking

Another area of weakness in traditional plans is the misconception around multitasking. People tend to pride themselves for being able to multitask; however, multitasking hurts productivity. When working on one task, there is a chance that a bottleneck or a delay in the task will get the resource to stall. Therefore, with two tasks, productivity improves as the resource can switch to one of the tasks while the other is blocked. With that said, people rarely have just two tasks and productivity rapidly drops off after more than two tasks are assigned concurrently. In addition to the loss of time due to the switching between different tasks (or projects), a person loses focus and concentration and it takes them even longer to get back to the level of focus they had prior to the interruption or the task shift. Figure 6.3 shows the decline in productivity when attempting to work on more than two tasks simultaneously.

Multitasking becomes an even more potent issue once some tasks finish late (Murphy's Law) because now the resources may report to the project on time, but they will have nothing to do—which, in turn, is going to further delay the project. An attempt to combat these issues often takes place in the form of attempts to increase overall team member utilization, but this will not help, and instead there should be an effort to maintain sufficient slack to cope with typical project variability. Loading team members to 100% capacity will not yield the expected increase in productivity. In fact, loading the team to 100% capacity is like loading a highway to 100% capacity—no one makes any progress.

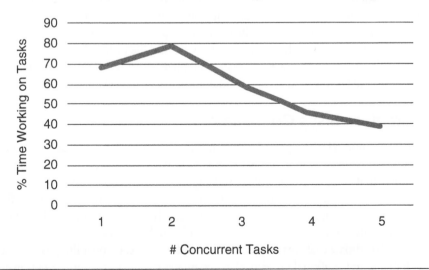

Effects of Multitasking on Productivity

Figure 6.3   Multitasking: the law of diminishing returns applies since productivity goes down when attempting to juggle too many tasks simultaneously

Feature Prioritization

Another issue that plagues traditional plans from being realistic is the lack of proper prioritization of features. Since waterfall projects are not scheduled to deliver value until the end of the project, there is less focus on the prioritization of features since it does not matter so much (especially not for the client) when features will be built. In many waterfall projects, the prioritization process is so lacking that it essentially does not exist and all features are of the same priority (which is often high). The only prioritization that may take place leads to the sequencing of work items based on the convenience of the development team, and with little to no regard to business value and customer needs. The rationale is that there is no need to prioritize features since all work needs to be completed at one point in time and the sponsor does not have any particular preference about the sequencing of the work.

As a result, toward the end of the project the team realizes it has fallen behind and it scrambles to meet the schedule. Since it is late in the game, the most common and natural course of action is to attempt to drop features, but with no effort to prioritize features up front, critical features may end up being dropped.

Uncertainty

In traditional plans, team members (and stakeholders) assume that the requirements analysis that was performed early in the project led to a complete and perfect set of specifications. Another prevailing and common assumption is that the sponsor will not change his or her mind, will not refine the options and opinions, and will not come up with new needs during the project. It is convenient to plan while ignoring uncertainty; and throughout the project many teams continue to ignore uncertainty about how exactly they will build the product. The tendencies around the ignoring of uncertainties are aligned with the practice of assigning precise estimates to the work that is inherently full of uncertainties.

Estimates versus Commitments

Finally, traditional plans provide estimates that attempt to be precise and accurate, but in reality, fail to be so. No plan can provide guarantees or even levels of accuracy that are going to satisfy the stakeholders' appetite for accuracy. The only way we will know the exact project costs for sure is to look at the actuals when the project is *over*. People cannot see the future and therefore we estimate based on our experience and on the information at hand—including its inherent uncertainties. Each estimate that we produce has a confidence level (probability) associated with it; and in traditional projects, these estimates get seen as commitments even though they are far from a sure thing. Figure 6.4 shows that the amount of uncertainty goes down with time as we make more progress in the project.

Reduction in Uncertainty

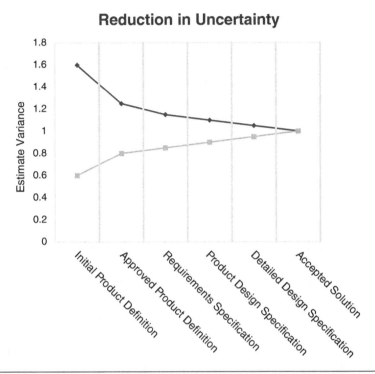

Figure 6.4   Reduction in uncertainty: the amount of uncertainty goes down as we make progress throughout the project; however, we cannot provide 100% accuracy until the very end. Due to the long planning horizon in traditional projects, they are less likely to produce accurate estimates for the long run.

THE BA IN TRADITIONAL PROJECTS

The role of the BA in traditional projects has evolved over the years. This chapter has covered a lot of the challenges that traditional project environments introduce and many of these challenges can be addressed and partially relieved by a BA. However, before moving on to cover agile planning and the role of the BA in agile projects, it is important to note a fundamental difference between the expectations about the BA role as part of the planning between traditional and agile projects. In traditional projects, the BA is there to compensate for the lack of planning and estimating skills among team members. In addition, the BA covers for challenges in communication, siloed mentality, lack of checks and balances, and inconsistent planning and estimating practices. Clearly, it is not the fault of waterfall/traditional projects—but the setup of these projects allows for all sorts of bad behaviors, lack of coordination, breakdowns in communication, and inconsistencies to exist. These bad behaviors are a result of the long planning horizons, the lack of ability to show tangible progress, the constant change, and the attempts to finalize specifications, requirements, and estimates too early in the project.

The expectations of the BA role as part of agile planning and estimating processes are different; they are more about complementing the already existing business analysis skills among team members, enhancing any areas of weakness, and ensuring that items do not fall between the cracks. The next section covers the areas of agile planning and estimating in detail, along with opportunities for the BA to enhance the process and add value.

AGILE PLANNING

Planning is an activity that in and of itself delivers value. Agile methods focus more on the value delivered by the planning activity than the plan it produces. Since we deal with changing environments, the focus should not be so much on the document that the planning activities produce (i.e., the project plan), but rather on the value of the planning activities. It involves producing timely information that is based on the most recent available knowledge. Further, agile planning is about understanding the actual performance, progress, context, and latest customer needs—and then acting upon this information—instead of spending precious time updating a plan. Agile planning is cyclical and happens throughout the project, not just at the start like waterfall environments where as the team makes progress and the needs change, the team must go through a risky and pricey change control process. There is planning early on an agile project, mainly at a high level, and then there are pulses, or cycles, of detailed plans—one iteration at a time—incorporating the latest set of information. By going through these planning cycles, agile planning balances the planning effort and investment against the likelihood that the plan will need to be revised throughout the project. As part of accommodating change, agile plans are designed to be changed easily in order to incorporate new knowledge and ensure the customer's latest needs are taken into consideration.

Agile Approach to Planning

Agile methods are designed to overcome many of the challenges introduced by traditional projects (i.e., projects that follow a deterministic approach). As such, many of the agile planning practices focus on doing things in a different fashion than the way they are performed in waterfall projects. Thus, they inherently reduce the need for a BA on the team since, in many traditional environments, the BA is in place to pick up gaps and deficiencies among team members. With that said, there are many areas where agile BAs can add value and enhance the already improved agile planning effort. Agile planning utilizes a multidisciplinary team to ensure that all points of view are considered and cover all aspects of the work required in the project—from requirements and design, to development, testing, and any other functions that are involved. This helps ensure that the planning is done consistently and with less unknowns so it does not miss any key elements. In addition, the planning process includes the sponsor or product owner because he or she plays a critical role throughout the planning process. On occasions when the product owner is not available, the BA can partially cover for this absence, but only to a degree. Generally, the BA should enhance the communication and the flow of information to and from the product owner.

Three Levels of Planning

The agile project is broken down primarily into three levels of planning: release planning, iteration planning, and day (daily) planning. This systematic breakdown within the project allows the team to perform planning activities in a focused manner where every level of planning provides more details that refine

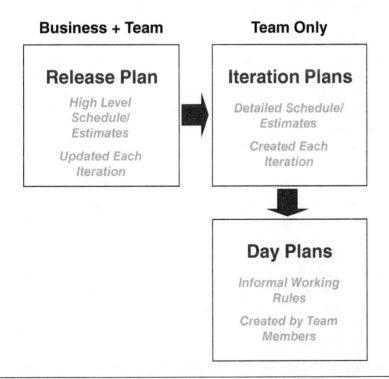

Figure 6.5   Three levels of planning: release, iteration, and daily. The three levels feed into each other and provide checks and balances for one another.

the higher level planning. Figure 6.5 helps with understanding the three levels of planning within the project and the system they provide for refinement and checks and balances. The project backlog is broken down at the beginning of the project into iterations and releases. Iterations are time-boxes that produce something of value each iteration (called product increment) and they are grouped together into releases. While iterations have a fixed duration that is set up for the project; a release is made up of one or more (typically up to three) iterations, and therefore there is no pre-set duration for a release. The number of releases and the release timing are determined by the needs of the customer.

The planning process for the project starts by breaking down the backlog into iterations and releases. The team determines its velocity so they can slice the project backlog based on the business needs that the product owner provides. The velocity determines the number of features to be produced in each iteration and the planning at this point is done on a high level to figure out the breakdown of the project backlog. Next, the iterations are grouped into releases to deliver the value to the customer. Although each iteration produces value, each release needs to have a certain level of critical mass to justify the process and cost associated with actually releasing the product increment to the client, and for the client to test it and subsequently use it.

Level 1: Release Planning

The first level of planning release planning is performed by the business (specifically, the product owner) and the team, where the product owner provides priorities and the team refines them and produces high-level estimates. The result of the release planning is a high-level schedule and estimates that are updated each iteration.

The release planning process focuses on building a high-level project schedule by identifying the following items:

- The number of iterations and the start and end dates of each iteration
- Releases and the start and end dates of each release
- Key business milestones (for the most part, these are provided to the team by the product owner based on customer needs)
- Any dependencies and interlocks with other projects
- Initial contents for each iteration, including iteration goals, themes, stories, and features; at this point, it is a high-level plan and therefore it does not contain a task—there is no point in planning on the task level since the actual progress, as well as the needs, are likely to change as the project moves forward

Figure 6.6 provides an overview of the release plan process and activities. Two primary considerations feed into the release planning process, including identification of high-level features and user stories, as well as considerations around

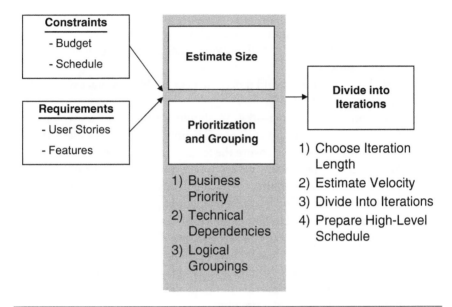

Figure 6.6   The release planning process: high-level planning leading to a cycle of detailed planning

schedule and budget constraints. At this point, the user stories and features are part of the project backlog, and they are not part of any order or structure. Next, the team estimates the size of the stories (we will review the estimating process later in this chapter), and the prioritization and grouping process takes place based on the following considerations:

1. Business priority
2. Technical dependencies
3. Logical groupings

After the prioritization and grouping process, the team moves to dividing the stories into iterations by choosing the length of the iterations, estimating velocity, and preparing a high-level schedule.

The BA and Release Planning

The BA does not have any specific role in the release planning process but depending on the needs of the project and the team, there are several potential areas for the BA to add value:

*   Help facilitate discussions
*   Ensure high-level estimates are realistic (including supporting the planning process)
*   Challenge estimates or provide context about estimates

- Challenge priorities—providing context for the product owner about considerations related to business value
- Support the team's role in identifying other considerations in the prioritization process
- Help look for answers, and if needed, help establish contact with subject matter experts (SMEs)
- Focus on value creation—the BA should always have an eye for value (and for reducing waste) and therefore provide another point of view in the planning process to ensure the focus is always on value creation
- Finally, the release planning process involves two levels of planning: (1) the high-level planning where the BA can help maintain focus and reduce likely confusion between maintaining the high-level approach of the plan and the need to have these estimates realistic and (2) the second level of estimating where detailed estimates for a small sample of stories become more likely to require help from the BA in ensuring consistency, while taking all considerations into account concerning coordination and accuracy of the estimates (later in this chapter we will cover the need and purpose for the detailed estimating portion of release planning)

Level 2: Break Project into Short Iterations

A key feature of agile methods is the breakdown of the project into small iterations. These time-boxes (typically 2–4 weeks long, but it is not mandatory for the iterations to be confined to this range) are the focal point of the detailed planning effort as the high-level plans from the release planning are refined—one iteration at a time—on a rolling wave planning basis. Iterations deliver products that are of releasable (or a shippable) quality. This means that the product increment that is produced by the iteration does what it is supposed to and performs to the extent that it needs to; however, the product increment that is produced by the iteration is not actually released to the client. The quality and performance of the product increment are such that, should the client choose to, the product increment can be released and used and benefit can be realized. The iteration's product increment as well as the theme and features of each iteration (and the grouping into releases) focus on producing business priorities and the team delivers features (and not tasks) using user stories that were prioritized by the business.

As we make progress, we take new considerations into account based on the value the team produces, the actual progress, and any potential new and changing client needs. The feedback cycles of the releases and the iterations allow the team to update the plan to reflect actual experiences and the benefits of new knowledge. With the guidance of the product owner, the team adjusts priorities at the start of each iteration to maximize the return on investment.

The iteration produces detailed schedules and detailed estimates for the product increment to be built in each upcoming iteration. The planning process takes the high-level estimates and plans from the release plan for the upcoming iteration and refines them. This process repeats for each iteration and further refines the release plan to reflect the latest information about the progress and the client's needs.

The BA and the Iteration Plan

The iteration planning process is about coming up with detailed planning (bottom up) for all the user stories and features by breaking them into tasks and coming up with accurate estimates for the work required to produce them. Even teams whose members have business analysis skills often require support, coordination, and guidance from a BA to ensure that the process is facilitated properly, is consistent, and yields accurate estimates.

Level 3: Day Plans

The third level, or layer, of planning is the day plans. These are informal working rules created by team members for the work to be performed each day. This is further slicing estimates and plans for the iterations into the daily activities that each team member needs to perform. The daily plans are the most detailed plans because they change from one day to the next based on the actual progress and the events that take place within the iteration. One day at a time, the team makes progress and updates the next daily plan accordingly with the goal to deliver as much as possible of the value, theme, features, and scope of the iteration. One day at a time the team chugs forward, making progress toward completing the iteration goal on the way to delivering the release. Each level of planning provides refinement of the *outer* or higher level plan (e.g., the iteration plan is a detailed plan of one section of the release plan that is due to be performed in the upcoming iteration, validating the high-level planning that took place in the release plan in a rolling wave planning fashion). Each accomplishment, progress, or problem helps update the actual progress, expectations, and future higher level plans. Whatever happens, each day is another piece of the puzzle of the iteration of the release so all plans at all levels reflect actual progress and the latest set of needs and information.

The BA and the Day Plans

The day plans are less formal and are developed by each team member based on his or her own respective needs. With that said, there are some elements that need to be consistent across the entire team. Although there are no clear guidelines for the daily plans, the BA can help maintain the consistency, ensure coordination, and facilitate some of the details of these plans.

The BA's Role in Planning

The role of the BA in traditional planning is clear: there are so many issues that the BA needs to complement and balance the team members' unrealistic planning process. In agile planning, the role of the BA in planning is less critical—depending on the planning and business analysis skills that team members have. While it is expected that agile team members plan and estimate, and the agile processes empower them to do so while providing multiple checks and balances to ensure plans are realistic, team members may still lack sufficient planning and business analysis skills to create realistic plans. Despite the short-term planning horizons and the rolling wave planning in agile projects, many team members may still lack the relevant skills and others may bring with them bad planning practices from previous projects. The fact that it is an agile project does not provide us with a guarantee that everyone will do and act exactly as expected and to the extent that we need them to.

Release Planning

Although performed at a high level, planning, estimating, and prioritization activities take place during the release planning process and they are critical for the agile project success.

The Prioritization Process

Prioritization is an important concept, no matter what the context or the project life-cycle type is. Prioritization is about accountability because when we have to do something, we learn what needs to be done, what is at stake, and what it will take to do it—and then we sequence things and put them in order. If we had enough time to do everything we need to, there would not be a need for prioritization, but with time and resource constraints, and with conflicting needs and priorities, our capacity is limited. The need for prioritization exists throughout the project, but it becomes even more important to address once we reach our capacity. At that point, prioritization is about making choices that maximize the value for our stakeholders—we can pick our battles, cut our losses, and stand behind our decisions. Once we reach capacity, which is the limit of what we can do (at a given time, context, or bandwidth), we need criteria to determine what we should move forward with first, what can be dealt with later, and what items can be postponed, reduced, or removed altogether. By not completing certain things we need to, we will make certain stakeholders unhappy, but the role of prioritization is to minimize the negative impact of not being able to achieve any one of our goals.

Agile methods are big on prioritization. This means that our project backlog turns into a prioritized list of items to do, and in turn, we slice these items and group them into iterations and releases. Why are agile methods big on prioritization? Once the team identifies its capacity to perform work, the prioritization process allows us to pick whatever adds more value and perform it within our capacity (team size, speed or velocity, and iteration length). It also supports the notion of maximizing the value we produce as we start by performing the important (and complex) items on the backlog and gradually move down the list. Further, the prioritization process must be clear, transparent, and consistent; it is all too common to see priorities change (in projects, personal life, or the public domain) with almost no notice, no reason, and no consistency—in an almost arbitrary way. Agile methods provide clear guidelines for prioritization and these allow us not only to prioritize items, but also to change the prioritization as we go based on performance, changing priorities, and other needs. The prioritization process is one of the main things that make agile projects pragmatic and adaptive.

The prioritization process also comes to support the product vision and the strategic goals. The main idea is that we need to produce business value, and in order to do so, we approach each user story in a consistent manner, checking the value it adds to the overall project/product success and incorporating additional considerations to ensure that this story is indeed the right thing to do next. Although implied, it is important to note that the prioritization process is about relativity; we take a list of user stories, requirements, or features and determine which one should be performed next in relation to the full scope of work that we need to deliver. There are four criteria for prioritizing stories and other work in agile projects: business value, technical dependencies, minimal feature set/product viability, and logical groupings. While the team, the BA, other stakeholders, and SMEs have a say in the prioritization process, this *say* is in the level of recommendations since it is ultimately up to the product owner to determine the order and priority of the backlog items.

1. Business Value

The first and most important prioritization consideration is business value, or business priority. It is provided to the team by the product owner and the team will work with the product owner if there is any need to make adjustments to any of the other items. The product owner, with potential help from others including the BA, determines the business value based on the following considerations:

- Financial value (e.g., net present value, internal rate of return, payback) or measures of customer service value
- Cost of developing and supporting the items under consideration (one way to determine this cost is to divide actual costs for the last iteration by the number of story points for an average cost per story point)

- Amount of learning required about the product and the process, as well as skill development
- Amount of risk removed
- Legal and compliance requirements

The business priority *calculations* are performed by the product owner informally. The process starts by ranking stories by value-to-cost ratio with the highest ones at the top of the list. The product owner then uses other factors (such as risk and learning) to adjust items up or down the list. The resulting sequencing will be to perform the high-value, high-risks items first, then move to the low-hanging fruit (high-value, low-risk) and prioritize the low-value, low-risk items last. Items that are high risk and low value should not be performed altogether. It is common that the product owner (and other team members) forgets to include non-functional requirements (e.g., security, user interface design) and the BA should be on guard for such items.

2. Technical Dependencies

Quite often the business value does not take into account any technical considerations. It is not expected that the product owner or the customer will fully understand all the technical considerations and this is where the team and the BA make their contribution to the process. Technical considerations are unique to the product, domain, circumstances, and organization. While these considerations do not change the overall business value, there is a need to adjust the order and sequence of things to allow value to be produced based on realistic and technical conditions. When someone buys a custom home, he or she may tell the builder that their main focus is on the second floor, and therefore they want the master suite to be built first. At this point, technical considerations must be taken into account since it is not possible to build the second floor first. As the product owner may not be aware of all technical dependencies and considerations, the team will provide this information to the product owner in an effort to get the product owner to change the backlog items' prioritization. The BA does not have a formal role at this stage except for the need to ensure that the right information is communicated from the team to the product owner and that all technical considerations are taken into account.

3. Minimum Feature Set (MFS) or Product Viability

This item may not be a stand-alone consideration in the prioritization process; it may fall under the fourth consideration—logical groupings. MFS or product viability is about ensuring that key functional (or non-functional) considerations are taken into account. Such items may be vital for the product to be considered successful, but they may not be at the top of a stakeholder's (including the product owner's) mind. For example, when a financial institution wants to launch a mobile loan application app, the key features of the solution are the main focus; but there are other items such as compliance or security that may be forgotten by some stakeholders. The BA does not necessarily have knowledge about all of these items, but he or she can ensure that there is a sufficient awareness level, provide checklists, seek advice from SMEs, or simply have another set of eyes in the process to ensure no items are overlooked.

4. Logical Groupings

This is where the team provides the product owner with additional considerations that may lead to a change in the overall prioritization of items. Logical groupings refer to anything else that is relevant—from resource allocation and availability, to focus areas, functionalities, and features that should be built within a different time frame than initially thought due to any one of these considerations. For example, if a resource with specific skills works on one feature, it may make sense to get this person to complete another, related feature even if the other feature is not of a high priority.

ESTIMATING COMPLEXITY

Before we go over the process of estimating the complexity of items, it is important to maintain focus and cap the amount of time the team spends on each item. No matter how much time is spent estimating, it is impossible to get to 100% accuracy and it is important to recognize that there will still be uncertainty. Most sponsors and managers request high levels of accuracy in the estimates—requiring large efforts, but with diminishing returns of accuracy. Agile teams focus on *good enough* estimates and try to optimize their estimating efforts. There is such a thing as spending too much time estimating and Figure 6.7 illustrates that point.

During the release planning process, the product owner provides the team with a list of items and the prioritization, planning, and estimating process begins. In order to estimate complexity, agile methods utilize consistent and effective tools that help the team estimate the level of complexity for each item, as well as consider each item's complexity in relation to other items. There are two ways to perform these estimates—ideal time and story points.

Ideal Time

Ideal time is similar to estimating effort and it refers to how long it should take a resource to perform an activity if he or she were dedicated to doing it. Ideal

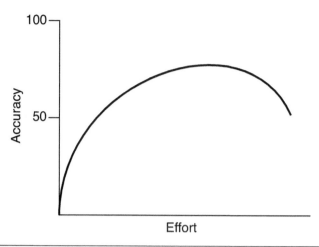

Figure 6.7   The law of diminishing returns: there is such a thing as too much planning

time differs from elapsed time because of the natural overhead and distractions we experience with activities, such as supporting the current release, sick time, meetings demos, administrative activities, multitasking, walkthroughs, e-mails, calls, and training.

Since there are often problems when estimates are given in *days* with no qualification as to what a *day* means, the term *ideal days* helps to relieve this situation because it is a measure of size—not duration. The team needs to be able to assign one estimate to a story that is the total of the ideal days estimate for each role required to complete the story, including management, analysis, design, development, and testing. The BA can benefit the team in these situations by ensuring that all aspects of a story are taken into account. Ideal days provide more in-depth analysis of the tasks required to build a story and they are easier to explain to outsiders, but there is always a risk that some stakeholders confuse ideal days with calendar days. While ideal days are easier than story points to estimate at first; skill levels affect ideal days estimates and, therefore, one person's ideal days may not be the same as another's.

Story Points

Using story points is more common among agile team members in estimating size based on a story's complexity. Story points are an abstract unit that is not directly convertible to hours or days and is based on the completion of the story or feature by the whole team. There are several schools of thought on how to estimate with story points, including the use of T-shirt sizes, a scale of 1 to 10, or a scale based on a Fibonacci series. There are pluses and minuses for each of the techniques:

- A scale that is based on T-shirt size is not quantifiable enough
- A scale of 1 to 10 does not emphasize enough the increasing gap in the sizes of the higher values
- A Fibonacci series is a sequence of numbers where each number is the sum of the previous two numbers (see Figure 6.8). The team assigns story points for each requirement only using the numbers on the cards. This technique allows the team to get an idea of how much bigger a requirement is relative to others. Notice as the numbers progress, the distance (effort) between the numbers increases as well. The team might make a rule that only user stories with an estimate of 8 or below can be completed in one sprint (or iteration). Along those same lines, anything from 13 or above would need to be broken down further before it is considered for one sprint.

Planning Poker

This technique uses the Fibonacci series cards and starts with the facilitator explaining the user story and related notes. Then team members ask the product owner high-level questions about the requirement. After the questions are answered, everyone selects a card with the number of story points they think the user story would encompass and puts it face down on the table. On the facilitator's cue, everyone turns over their card. If there is a big range of estimates, the team discusses why to make sure that no one misses any piece of

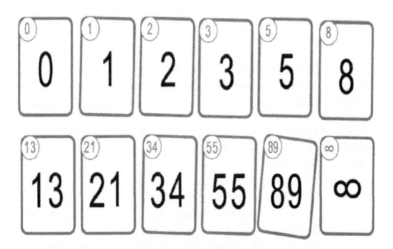

Figure 6.8   The Fibonacci series: playing cards as part of estimating size in agile projects makes the process effective and helps in team building

important information. Based on the processes the team has, there will then be either another vote on the number of points necessary to complete the user story or a discussion that leads to a consensus-based estimate. The planning poker process ends when consensus is reached or when the range is minimal and explainable.

The BA can take a role that supports the team and the Scrum Master/PM by helping the facilitation process, challenging estimates, looking for answers where there is uncertainty, helping the discussion process, and helping individuals come up with realistic estimates:

- Assign the smallest user story a value of 1.
- Then, find the next largest item and assign it a 2.
- Using those as reference points, estimate the remaining smaller user stories, scaling them to the starting pair (e.g., a story that is a 4 should be twice as complex as a story that is a 2). The progression into gradually larger stories (two to three times in size) is proven to be very effective and intuitive for people since multiplying by larger numbers—although mathematically possible—introduces difficulties for people to comprehend when discussing story size. It is important to not overanalyze: the team needs to ask the product owner a few questions, make some assumptions, estimate, and move on to the next story. The BA is useful here in ensuring the team is not overanalyzing, providing context to and from the product owner, and introducing and documenting the assumptions.
- Once the small ones are estimated, move on to the medium-sized ones and estimate those, leaving the very large ones to the end.
- The team should not cap the number of story points assigned to a story and should keep going as long as there is a need. Typically, at this stage of the release planning, the team will likely reach a level of 34 to 55 points for its largest stories.

Finally, and with the help of the BA, the team needs to apply some adjustment factors to the estimates in order to account for all sorts of factors that may introduce variability to the story that was estimated. Those factors may include:

- Poorly conceived options (which could be discouraged by adding some points that can help stakeholders realize that a story may not be a good idea to pursue after all)
- Unclear or fuzzy requirements
- Lack of stakeholder buy-in for the project
- Resource availability issues, especially when the team is not fully dedicated
- Large or distributed teams
- Work environments are dysfunctional and there are communication issues
- Technical or business risk

It is important that this process is consistently done in a controlled way and that the team does not get too granular or heavy-handed with these adjustment factors.

The result of the release planning process can look like the table provided in Figure 6.9.

Book Writing Project

Project Backlog

Feature ID	Description	Business Priority	Dependencies	Complexity (in Points)
20	Book Marketing Plan	H		7
21	Back Cover Marketing Blurb	H		1
3	Introduction	H		2
4	Chapter on Agile Requirement Gathering	H		5
5	Chapter on Agile Estimating	H		6
6	Chapter on Agile Planning	H	5	7
7	Chapter on Backlog Maintenance/Replanning	H	5,6	2
9	Chapter on Daily Meetings	H		3
10	Chapter on Velocity	H		6
11	Chapter on Reporting	H	10	4
23	List of Possible Early Endorsers	M		1
19	Solicitation Package for Early Endorsements	M	23	3
22	List of Final Book Reviewers	M		2
24	Book Review Package	M	22	3
16	About the Authors	M		1
8	Chapter on Agile Change Management	M		6
13	Chapter on Agile Testing	M		5
1	Acknowledgments	L		1
12	Chapter on Dealing with Stakeholders	L		4
14	Conclusion	L		2
17	Index	L	3,4,5,6,7,8,9,10,11,12,13,14	10
18	Additional Resources	L		2
15	Bibliography	L		1
2	Dedication	L		0
			TOTAL	84

Figure 6.9   The project backlog: in this example of writing a book, the project backlog can look like this—for each item there will be a priority reading, dependency the story has on other stories, and a complexity rating in points. The complexity in this example is measured in points from 1 to 10. Note the numbering sequence on the left as the original order of the stories changed to accommodate for each story's priority

VELOCITY

Now that the team has a backlog to work with, there is a need to slice the backlog into iterations. We start by choosing an iteration's length: many environments have a set length for all their agile projects' iterations, which is commonly between 2 and 4 weeks. There is no *magic number* for an iterations' length, but it has to be consistent throughout the project with the following considerations:
Factors to consider:

- The nature of the work and the need for feedback
- The size of features
- The team's experience level
- Available automation tools

Another important factor in determining the iteration's length is velocity. Velocity is a measure of the team's rate of progress and it is calculated by summing up the number of story points or ideal days that the team completed in its most recent iteration. For example, if a team completed ten story points in the last iteration, we can assume they can do ten again in this iteration. Agile methods estimate size, not duration—and duration is derived from the size. Once we have an understanding of the team's velocity, we turn the story points or ideal days into a schedule by applying velocity. We add up the total story points for the project to get the team's velocity for a specific iteration length. We then divide the points total by the velocity to determine the number of required iterations— and this forms the basis for the release plan. The BA is often involved in this stage since he or she consolidates the estimating efforts and ensures that the process remains consistent and does not lose integrity due to the many considerations, stories, team members, and other moving parts.

One of the great aspects of using velocity as a measure for planning is that it does not matter if our estimates are correct as long as they are consistent. With consistent estimates, measuring velocity over the first few iterations will allow the team to prepare a more reliable schedule downstream. To estimate the velocity of new projects, we can use one of the following methods:

- *Historical values*: this is applicable only if the team, tools, and technology are the same
- *Run an iteration to get actuals*: after three iterations, we can proceed forward based on the average or range of our actual performance in these iterations
- *Make a forecast*

Forecasting Velocity

The following is a basic approach for estimating velocity and its main weaknesses are that it assumes that all team members who are included in the calculation are equally productive and equally producing deliverables; it therefore does not identify overburdened resources:

1. Estimate the number of hours per day available to work on the project for each team member (typically, 60–70% productivity for dedicated resources).
2. Select the number of team members to be used—only those producing deliverables, not managers, BAs, etc.
3. Select an iteration length.
4. Determine the total number of hours spent on the project during an iteration (# of hrs per day × # of people on team × # of days in iteration).
5. Pick a few random stories and expand them into tasks and estimate the number of effort hours per task, getting a total number of hours for each story. This is about going through a detailed planning process (bottom-up) for these stories. Even though agile methods call for planning only for what has to be performed next, selecting these medium-sized stories and estimating them in detail will help benchmark our high-level story point estimates and extrapolate from the detailed estimated stories to the rest of the stories. This takes us back to the need to be consistent in our estimates—as with consistent estimates, all we need is roughly six sample stories to estimate in detail.
6. Gather enough stories for the iteration so that the number of hours for each story adds up to the number of hours available in an iteration.
7. Count up the story points for the selected stories to determine projected velocity.

Necessary adjustments will need to take place for teams that do not have its members equally utilized and productive.

Next, using the forecast velocity, we carve up the project backlog into chunks of the appropriate size (give or take a point or two is generally ok). If a chunk is much smaller than the velocity forecast, but the next piece would push it way over, we need to look down the list for the next piece that would fit and move it up the list into the current chunk of work. Each chunk represents the scope for one iteration.

Before moving forward to the iteration planning, there is one more thing to do: plot the iterations on a calendar with real project start dates in order to get a realistic view. Otherwise, we may end up with a project backlog that tells us how long a project will take to complete, and although these are realistic estimates, it will fail to deliver on the promised date if it did not consider all calendar constraints along the way. For example, if a five-month long project is scheduled to start on October 1, we need to consider the following:

1. *Actual start date*—The fact that we tell our stakeholders we can start on October 1 and it will take five months does not mean that we get

the go-ahead on time. It is common to see projects where stakeholders believe work is progressing when no approvals have been provided. By the time there is permission to move forward, it is already well into November. The team still needs five months, and the stakeholders may think that we are already into week five of a project that we haven't even started.

2. *Periods of lower productivity*—Every organization has down periods that we need to account for; but in every environment, there will be two periods of lower productivity: summer and over the Christmas holidays. As a result, if the project is scheduled for five months, at the very least, we cannot count the two weeks leading from mid-December to the new year as a fully productive period. Even if there are no shutdowns over the holidays, a large number of team members will miss work over this period and it will slow the project down. For the iteration(s) that cross over the holidays, we should just set expectations for a lower productivity (or velocity).

In addition, even though we believe the project can deliver within five months, we need to allow for complexity and risks. On high-uncertainty/high-risk projects where the end date cannot move, it is a good practice to leave one or two iterations before the end of the project to absorb things like scope increases, the risk of technology challenges, and defect correction work. If we plan for two-week-long iterations over five months, there should be 10 iterations, but we should really plan for eight or nine iterations that are populated with stories with the last one or two (depending on the amount of uncertainty) to be left empty to absorb those items discussed previously. This is a form of a buffer, but instead of asking for more time we leave the last iteration or so open.

Once we have the project backlog carved into releases and iterations, there is one more set of adjustments to make by moving stories and features from one iteration to another to reflect dependencies, key business milestones, constraint dates or periods, lower productivity periods due to planned holidays, resource leveling, and any other concerns around the required release functionality. Next, we move to perform the iteration planning process one iteration at a time.

A Word About Change

Just because there is a closer working relationship with the sponsor or product owner, and there is more flexibility in the project overall, it does not mean that the project is out of control. Formal change management processes still need to be implemented, including formal sign off on adjusted schedules, budget, feature lists, and resource allocations. Usually, all changes negotiated at the end of an iteration are grouped onto one single *omnibus* change authorization form, and the signed change authorization provides the mandate for the new project backlog, release plan, and upcoming iteration plan. Sometimes this requires multiple signatures as multiple stakeholders are affected; however, this consolidated approach keeps the paperwork to a minimum. Due to the timelines to perform these changes, it is often the BA who facilitates the change process, documents it, and ensures that all considerations, SMEs, and stakeholders have been included in it.

Iteration Planning

The iteration planning process *walks the talk* of the concept of *rolling wave planning* and is essentially about verifying the velocity assumptions that are made during the release plan. Planning for an iteration involves taking the chunk of work that was identified and confirmed for the upcoming iteration and performing traditional planning (detailed, bottom-up)—but only one iteration at a time. It involves developing a work breakdown structure—identifying the task, the task dependencies, and the task owners for the iteration. Then the owners estimate the work effort in hours (or days, but more likely and more accurately in hours). The outputs of the iteration planning process include a project schedule, the iteration backlog, and the initial iteration burndown (tracking) chart. While during Iteration 0 we perform the high-level estimates for the release, toward the end of Iteration 0, we go through a detailed round of planning for the work to be performed in Iteration 1, as shown in Figure 6.10. We then continue

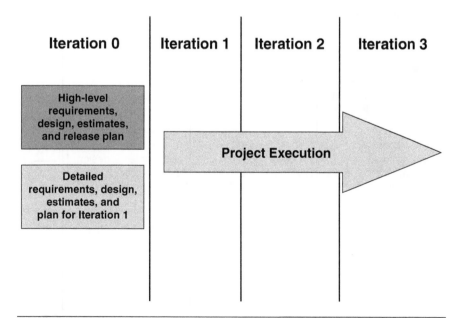

Figure 6.10 Check top-down estimates with bottom-up planning—one iteration at a time

the process, one iteration at a time, to verify the high-level estimates from the release planning for the upcoming iteration. This is a form of fast tracking and the BA can play a key role in this process: working with the design team members on requirements and verifying estimates one iteration in advance before supporting the team on their current iteration work. Figure 6.11 illustrates the concept of fast tracking. The detailed planning is done one iteration in advance while the team is working on the current iteration's work that was planned, in detail, during the previous iteration.

Some would argue that planning one iteration in advance and before the team gets the formal feedback during the demo is risky. While fast tracking techniques are considered to be risky, it is a calculated and responsible risk. By fast tracking, we do things in parallel that otherwise we would perform in sequence. While it is inherently risky, the team's performance, issues, and progress are known at this point and the team, with the help of the BA, can estimate the activities for the next iteration with a fairly high level of confidence. While there may be some changes as a result of changing needs or performance issues leading to the demo, it is unlikely that the team enters the demo with a completely different idea of how they are performing than what is reality. Since the iteration is a short planning horizon, we already have the high-level plans, have verified the plans, and know our performance levels, so we are looking at a low level of

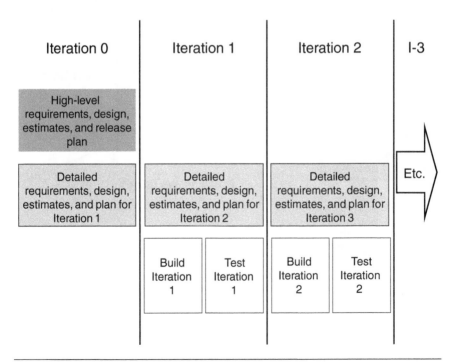

Figure 6.11   Fast tracking: detailed planning—one iteration at a time

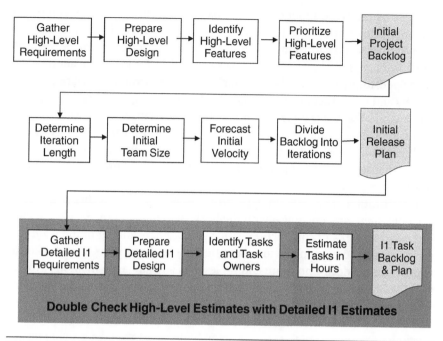

Figure 6.12   Each iteration contains Iteration 0 of the next one

risk that produces great benefits and saves time for the project team. Figure 6.12 shows the process that the team goes through in Iteration 0 and then the shift from the high-level activities to the latter part of Iteration 0—where detailed planning activities take place for Iteration 1. In a way, we can say that at some point, every iteration becomes the Iteration 0 of the next one.

RECAP

There are many differences between agile planning and estimating and the planning practices in deterministic and waterfall projects. Many of the agile planning practices are used to overcome challenges that are introduced as a part of waterfall projects—so the team can produce estimates that are realistic and reflective of the latest set of needs. The problems that are associated with waterfall planning and estimating practices are not necessarily the *fault* of the methodology, but rather they are due to practices that lack sufficient checks and balances, planning horizons that are too long, and unrealistic expectations for accuracy due to changing and unclear conditions.

Agile methods are about checks and balances. It almost appears as if the lu-minaries who came up with the concepts that make up agile had teams that they did not trust, so they introduced mechanisms to almost micromanage the team, along with *loops* of planning that protect each other, feed into one another, and offer opportunities to take the early high-level plans from the release planning process and refine them with pulses of detailed planning one iteration at a time. This ensures that the team responds and adapts to change, works with the latest set of information, and incorporates the changing priorities and stakeholder feedback into the planning process. There are also techniques (e.g., planning poker) that serve as team-building exercises that also help produce realistic and consistent estimates.

Prioritization is also key in agile planning. It is a process that is transpar-ent, consistent, and ongoing—with the buck stopping with the product owner. The process works in conjunction with a solid understanding of the team's ca-pacity, measured by velocity, to ensure that the team commits to and takes on what it can produce—no more and no less. As a result, agile methods provide a strong ability to produce plans that are consistent and that are verified in a rolling-wave-planning fashion.

The BA, while having no formal role in the planning process, has ample op-portunity to provide support to the team. Some areas include: provide insight into estimates; seek information from SMEs; confirm requirements and details; address gaps, assumptions, and constraints; and work with the product owner. The BA can add significant value in one additional area by establishing produc-tive working relations with the product owner—but not by simply providing support for the product owner, *replacing* the product owner, or doing the same old role that BAs do (covering cracks, gaps, and competency issues within the team). The new role for the BA utilizes skills that BAs already have, but in a different way: by enhancing the role of the BA through helping the product owner focus on *adding value*. This concept can help agile projects not only de-liver more of the intended value—and possibly faster—but it can also increase the overall amount of value the project provides. The BA role is about engaging the users in such a way that helps them see beyond their identified needs; ex-panding the possible amount of value that the project can produce.

In short, agile planning works for a series of reasons that all work together toward producing realistic plans:

- Re-planning occurs frequently
- Estimates of size and duration are separated
- Plans are made at different levels
- Release plans are based on features, not tasks
- Work in progress is eliminated in every iteration
- Uncertainty is acknowledged and planned for

Finally, the argument about whether agile projects even need BAs is legitimate. Agile planning and estimating practices cancel more of the need for *traditional* BAs; however, BAs can really transform the contribution they make to a value-based approach where the BA is there to enhance the contribution of the team members to the planning and estimating process—in areas that are applicable for the situation. This means less of the traditional *translating* role, where the BA is simply there to cover for shortfalls in the process, and more as an enhancer or facilitator of efficiencies with an eye for value and waste reduction. The agile settings bring out the original potential of the BA, and with a proper mandate and collaboration with the product owner, the BA can unleash the potential to produce more value for the customer through producing more value for the team.

7

AGILE EVENTS AND
WORK PRODUCTS

Agile projects are known for a series of events, ceremonies, and work products that drive the agile process and help ensure delivery of value, collaboration, reporting, and transparency. Agile events and work products are essentially at the core of the benefits that agile offers and include the daily stand-up meeting, the project backlog (or as it is called in Scrum—the product backlog), the definition of *done*, demos, and retrospectives. These work products and events enable the management of change and ensure that value is continuously delivered and evaluated through reprioritization and backlog maintenance. This chapter provides a review of these events, explores ideas and techniques to ensure they are optimized for the project needs, and discusses the role of the business analyst (BA) in helping make these events successful, or the need for business analysis skills in doing so.

PRODUCT BACKLOG AND PRIORITIZATION

As there are some differences between the language that is used in agile project management and Scrum, we will continue to use the terms that align with agile project management. Similar to our use of the term *iteration*, rather than *sprint*, we will refer to the backlog as the *project backlog* and not *product backlog*.

The product owner creates and *owns* the backlog. The product owner brings in the business needs, determines the features and stories that will be included in each release and iteration, serves as the ultimate decision maker for prioritization of the items on the backlog, and makes all decisions that need to be made about the backlog. The project backlog is the scope of the project and, in turn— as it is sliced into smaller pieces—of each release and iteration. The timing of the initial backlog creation is at the very start of the project; a high-level list of items for the backlog is created, typically during Iteration 0. Although the product

198 Agile Business Analysis

owner is the owner of the backlog, at times the product owner's availability is not sufficient for the project needs and the product owner appoints a *proxy*, or some sort of a representative, to act on his or her behalf. Many environments have a BA *acting* as a product owner. We touched on this in previous chapters of this book, but it is important to reiterate that while the BA may be an *agent* of the product owner—providing context, streamlining communication, and improving the product owners' decision-making ability—the BA should not outright replace the product owner. The BA does not have enough organizational and senior-level context to do so effectively, and furthermore, the BA does not have the seniority or the capacity to replace the product owner.

Owning the backlog does not mean that the product owner acts alone; the backlog prioritization (and subsequently maintenance) activities are led by the product owner, but there is a need for community participation in backlog creation and in suggesting updates. By saying *community*, we refer to team members, subject matter experts (SMEs), and other stakeholders (as required) to ensure the backlog is prioritized and represents the most recent and relevant set of information. If there is a BA role in the project, this person can help identify and access those who need to provide information that helps the backlog building and prioritization process, as well as enhance the communication with these stakeholders. Depending on the technical level of the BA, he or she can provide some answers on behalf of the various contributors to the backlog.

The project backlog is a prioritized list, containing descriptions of all the features and functionality that are desired for the project's product to include. We can recall that in agile there is no need to elaborate and finalize all requirements up front; the product owner, with help from the team, will put together all the items that they can think of for the project on a high level and will add details for each story and functionality later as required. By saying later, it means that more information for each requirement will be required in a timely manner (just enough, just in time) for the process of building that requirement's product. Later means that during the early stages ambiguity is allowed for backlog items until information is required. Later there will be a better understanding of the solution (the big picture) and how the requirement fits in. Uncertainty will be resolved over time and that will translate into a reduction in the amount of time that would have been wasted while attempting an early understanding of requirements that may change or be canceled later. Here too, a BA can help look for information, identify candidate stakeholders for information, and streamline communication with them about the requirements. It is important, however, to keep in mind that there is a need for a high-level set of requirements early in the project in order to help you prepare an overall solution outline and model.

The early information that is captured in the initial backlog, along with a breakdown of the backlog into releases and iterations, is most likely sufficient for planning in detail for the first iteration. The backlog will then be allowed to grow and change as more information is discovered and more is learned about the product and the customer. The project backlog should contain deliverables and deliverable features, as well as defects, technical work, and knowledge acquisition—with some examples listed here:

- *IT project*: customer software features, packaged software installs, servers, router configurations, maintenance manual, user manuals, and defect bundles
- *Consulting project*: key interviews, report chapters, draft report, final report
- *Construction project*: foundation, framing, mechanical, cladding, etc.

In addition, key business events that require significant preparation by the team or other capacity should also be included:

- Analyst presentations
- Trade shows
- Architecture review board presentations
- Project review board submissions

While requirements on the backlog can be expressed in many ways, the most common (and possibly effective) way to express the features is in the form of user stories. As user stories are typically used to express functional requirements, there is a need to ensure that non-functional requirements are also expressed and included in the backlog. Ideally, agile team members will also *act* as BAs and in this context ensure that non-functional requirements are included in the backlog; but if there is a BA in the team, he or she keeps an eye out to be sure that the non-functional requirements are in place.

The backlog includes the features that were identified at the start of the project (during the release planning), new needs that have emerged through this past iteration, changes to previously discovered needs, and defects that have not been fixed throughout the iteration and are now waiting to be fixed in a future iteration. All these items need to be included in the backlog.

Iteration Planning

The iteration planning on the first day of the iteration is a direct continuation of the activities performed on the last day of the previous iteration. Chapter 9 provides more context in this area, as well as a breakdown of the daily routine for the three *types* of days in the agile project environment—in the following order: the last day of an iteration, the first day of the next iteration, and any other day of the iteration. The product owner brings into the iteration planning the prioritized project backlog and describes the top items on the list to the team—focusing on the backlog items to be performed in this coming iteration.

The team, in turn, determines which items they can complete during the com-ing iteration (based on team velocity) and the team members then move items from the product backlog to the iteration backlog. While the project backlog is made up of features, functionalities, and deliverables, the iteration backlog is going to break down each story and deliverable into the tasks and activities to be performed during this iteration.

The team goes over the prioritized project backlog and moves down the list from the higher priority items until it reaches its capacity—capping the list with lower priority items from the project backlog to maximize the value to be pro-duced in the iteration based on the iteration's goals and theme. It is easy to see that with the absence of relevant business analysis skills among team members, a BA can help in the story selection, the breakdown of individual tasks, the es-timating process, and the team's identification of its capacity. The BA can also suggest certain candidate stories for the iteration based on additional consider-ations, such as bugs that need to be fixed, stakeholder needs that may not have been captured, and other organizational considerations.

The Importance of the Backlog

The backlog serves as the connection between the product owner who rep-resents the needs of the business or the customer and the development team. The product owner may reprioritize work in the backlog at any given point due to the changing needs of the customer and the project—including, as a result of customer feedback, changes to estimates or new needs that emerge. The changes to the backlog (discussed later in this chapter) take place *in between* iterations, as much as possible, in an effort to keep the interruptions to work in progress during the iteration to a minimum. Allowing the team to proceed with the iter-ation plan with no interruptions helps the team maintain focus and flow and supports the effort to keep the team's morale high.

People who are new to agile often struggle with the notion that while agile is about dealing with change and delivering on the customer's changing needs, there is no change (as much as possible) during the iteration. This means that upon completing the iteration planning, the scope for the iteration is *locked* for the duration of the iteration. Unless there are *game-changing* factors that intro-duce themselves during the iteration, the iteration backlog continues to guide the team's effort until the end of the iteration. With that said, it is reasonable to expect that due to the relatively short iteration's length (commonly 2–4 weeks), the scope of work identified for the iteration is valid and should remain until the iteration's end. With backlog maintenance and reprioritization taking place at the end of each iteration, the most up-to-date, recent, and relevant needs of the customer should be reflected in the iteration planning. The backlog mainte-nance also includes any necessary adjustments to the release plan (and, in turn, to future iterations) based on new needs, the actual work completed in the most recent iteration, changing needs, defects, and any other considerations that help deliver value for the customer.

As the backlog tends to grow, at some point it may reach a level that is beyond the team's capacity to deliver in the project's timeframe. At this point, the prioritization process (with guidance and help from a BA) should look into lower priority items and other issues and mark them out of scope. This helps ensure that the scope of work (the backlog) moving forward is manageable and within the team's ability to deliver. Another issue that takes place in many project environments is the absenteeism of the product owner. Often, the product owner prioritizes the backlog at the start of the project, but is not sufficiently involved later in the project and does not make the necessary adjustments to the backlog. This is likely to turn the backlog stale since it will no longer reflect the stakeholders' and the team's feedback, and it will not provide the necessary value for the customer. A BA can get involved, ensuring that the product owner adjusts the backlog to reflect the feedback, but a BA cannot completely replace the product owner in this process. Another area that often impacts agile teams is when the backlog is not shared sufficiently (or often enough) with stakeholders—preventing information and feedback from flowing. This area is easier to address by a BA who can ensure that the backlog is shared, that stakeholders are involved, and that information and feedback flow. The BA can also help the team ensure that the backlog contains both functional and non-functional items, but also that the backlog reflects customer-facing and other necessary items.

Project Backlog(s)?

The product owner needs to rigorously maintain the project backlog to ensure it is reliable and fully reflects the most recent and relevant customer needs for the project. In order to better organize the items on the backlog, it is recommended to *break down* the backlog. The breakdown is not in reference to *slicing* the backlog into releases and iterations, but in a different context. Figure 7.1 provides an illustration of this breakdown into three different backlogs: the project backlog, the shadow backlog, and the defect backlog. The purpose of breaking down the backlog is to ensure that the team and stakeholder are focusing on the *main* project backlog while at the same time keeping other needs in mind, even though they have not been prioritized into the *main* backlog as of yet.

Since the project backlog includes all the stories and requirements for the project—in order of priority and broken down into releases and iterations—there are often challenges regarding how and where to keep and track other needs. Instead of *parking* them at the bottom of the backlog as a *pile* of ideas that have not been considered, the shadow backlog serves as a *parking lot* for new ideas that have emerged throughout the project, as well as for changes that have

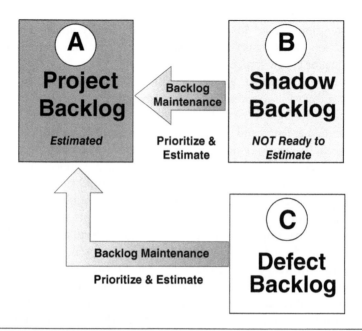

Figure 7.1   Project backlogs: breaking down the project backlog can make it easier to manage, prioritize, and maintain

been brought up. The problems with these items is that they are not for immediate implementation and they are not important or urgent enough to justify taking them through the process of prioritizing and estimating. So, until they become more relevant and timely, they will sit in the shadow backlog and wait for *if and when* it is their time to be considered. Should they become relevant, they will go through a backlog maintenance process where they will be prioritized (into an iteration, and in relation to other backlog items) and estimated so the team can perform the work pertaining to these items.

The same idea applies to the other backlog—the defect backlog. It includes the defects that are discovered through the iterations but not yet resolved—what is referred to as technical debt. These defects are going to be recorded in the defect backlog and wait for their time to be fixed. Here too, like in the shadow backlog, the defect backlog contains defects that are not important or urgent enough to be fixed or addressed immediately and they are *parked* in the defect backlog until they are ready to be prioritized and estimated as part of backlog maintenance. The rationale behind the separation of the shadow and the defect backlogs is to ensure that there is a separate tracking system for new needs and changes as opposed to defects.

If there is a BA on the team, he or she acts as the *manager* of the shadow and the defect backlogs. The BA logs the information in these backlogs as it is being brought up (throughout the iteration or at the demo); then the BA tracks these items and possibly introduces them for discussions as part of iteration planning depending on the iteration's theme and goals. It is not the BA's role to determine if items from the shadow and the defect backlog are going to be brought forward—prioritized and estimated—into the main backlog, but the BA can make suggestions and help explore the prospect of promoting these items by engaging the relevant stakeholders and helping with the process of investigating and assessing these items. Proper breakdown and management of the project backlog helps the team and the stakeholders maintain focus, channels everyone's effort toward the higher priority items, and provides a clear (and visual) separation of items in the backlog and items that are only parked there as candidates.

Once the project backlog is created, it remains a living document that is prioritized on an ongoing basis as part of the backlog maintenance process to ensure that the backlog fully reflects the latest set of information, knowledge, needs, and project performance.

END OF ITERATION DEMOS

At the end of the iteration, the team will demonstrate the product increment that was built during the iteration. The goal of this meeting is not the demonstration in and of itself, but rather the demo is an opportunity for the team to showcase the product that they built; and by that, show the customer and the other stakeholders the value that has been produced. The demo is a product review and the customer, along with the product owner, will make decisions as to whether and to what extent they accept each of the stories, requirements, and features that have been built in this past iteration. The demo is part of the iteration review process that is highlighted by the inspect and adapt process—a center stage concept in agile.

Based on the demo, the product owner will make a decision as to what backlog items are accepted (and moved to *complete*) and whether and to what extent deliverables need to change in order to become acceptable. The demo meeting is facilitated by the product owner, but in many environments, team members run the meeting. Ideally, with team members demonstrating the work they completed, the role of the BA in the demo portion is limited. However, the BA often assumes a major role in the process of obtaining feedback, ensuring it is clear, documenting it, clarifying which items go in which backlog, and what the backlog items for the upcoming iteration are going to be. Even high-performing teams, where team members successfully perform the role of the BA, often struggle in capturing and consolidating the stakeholders' feedback in the demo and translating it into a clear mandate for the following iteration. The BA can also take part in the demo setup and in the preparation and rehearsal process.

Let's review a few basic concepts about the demo:

- *Participants*: The demo is led by the product owner and the customer should be there. Participants include team members and internal and external stakeholders.
- *Inputs*: The idea of the demo is to review the progress and the product that was built during the demo against the plan. Therefore, we will have the iteration's theme and goals, the iteration's backlog, and the product increment that was just built during the iteration.
- *The process*: Each team member will demonstrate what he or she built in search of feedback from the stakeholders and the product owner. A short overview precedes the demonstration of each item (inspect), followed by discussion and feedback (adapt). This is the essence of the process with the goal of having clear understanding of what has been accomplished in the iterations, whether there are any issues and technical debt pending, and what will be the backlog for the next iteration. The feedback may also lead to changes in the release plan to reflect the latest set of information as articulated in the demo. The review process allows the product owner, stakeholders, and the team to learn about the delta between the work that the team committed to delivering and the work that was actually completed. The meeting may also cover, if required, key decisions that were made during the iteration, a review of project metrics, and a priority review for the next iteration.
- *Results*: It is desired that the customer and product owner produce immediate feedback for the items demonstrated so the team can proceed to the next iteration with a clear mandate. It becomes an issue when no clear feedback is produced. Overall, the feedback translates to backlog maintenance—which involves identifying items that go on the backlog (or the shadow/defect backlog), prioritizing and estimating, and updating the release plan.
- *Admins*: The duration of the demo varies based on the iteration's length and the number of features completed and demonstrated. However, in order to be realistic and respectful of the stakeholders' and the product owner's availability, the meeting should not last more than a couple of hours. Figure 7.2 shows a generic agenda for the demo and review meeting.
- *Prep*: Prior to the demo, the team should prepare for it and go through a play-by-play rehearsal. While it is not possible to fully anticipate the nature and extent of the feedback that stakeholders will produce, it is important to properly rehearse and make sure that the items that are going to be demonstrated are ready and working properly—and that the team members who will be demonstrating are also prepared. Demonstrating items that do not work as expected and having technical or presentation

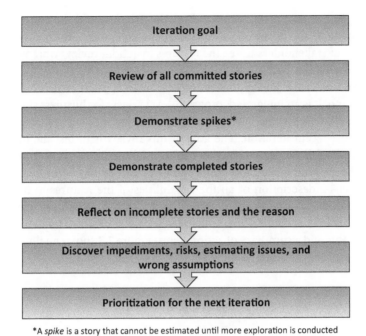

*A *spike* is a story that cannot be estimated until more exploration is conducted
by the development team.

Figure 7.2   Demo agenda: the purpose of the meeting is to present to the product owner and other stakeholders the completed work from this past iteration so we can go through the process of inspect and adapt.

issues during the demo is both disrespectful toward the stakeholders who are attending and is quite damaging from a trust-building and relationship perspective. The BA often takes a lead role in orchestrating the rehearsal and ensuring that the team is ready for the demo.

Different Types of Demos

There are up to three types of demos that an agile project goes through in the process of moving stories and features from their current state (which is still abstract) to actual performance:

- *Team demo*: The demo we described previously is the team demo where the team reviews the iteration's increment. The intent of the team demo is to trigger immediate feedback from the sponsor and the customer. This demo is done at regular intervals at the end of the iteration and is a Scrum-prescribed ceremony. It is one of the final acts of the iteration and it serves as the main measure of the team's progress. It is an opportunity for the team to show the progress and the contribution it made to the business and it allows the customer, product owner, and other stakeholders to see the features in action, review what works, and provide feedback.

- *System demo*: For IT environments, there is another demo—a system demo. It is an integrated, program-level review for business stakeholders of all the software delivered by all the teams (if this is applicable) in the most recent iteration. The goal of this review is to test the integration of the full system. It is a short session (about an hour) that takes place shortly after the iteration's end and involves a review of the business context—a description of all the capabilities of the features to be demonstrated—leading to the demo. At that point, stakeholders may present questions and the demo concludes with a review of risks and impediments and a summary of action items. Figure 7.3 shows an illustration of the process leading to the system demo.

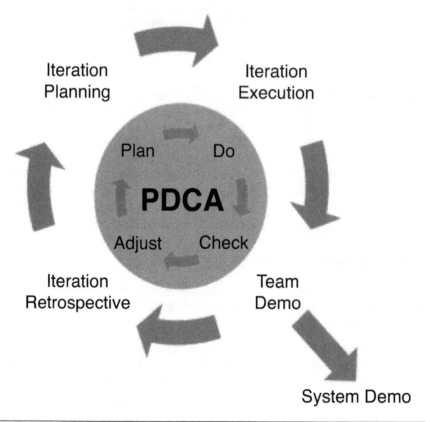

Figure 7.3   The process leading to system demo

- *Solution demo*: A third type of demo is the solution demo. In projects that utilize this type of demo, the solution demo can be viewed as the center of the program increment cycle where the results of all development efforts from multiple releases are made visible to the product owner, customer, and other stakeholders for feedback and observations. In a solution demo, the team demonstrates the solution's new capabilities, along with the solution's compliance with non-functional requirements. In a team demo, the focus is only on items that can be demonstrated, but here the team will also include non-functional items that are harder to demonstrate. The solution demo will also use the results of the most recent system demo, but the timing of this demo is different—it takes place after the system demo and therefore some time after the end of the iteration. The session should last no more than a couple of hours and it includes a review of the objectives, a demonstration of each objective through use cases, an identification of the business value that was completed for each objective, and a Q&A session followed by a summary.

Demo Challenges and Issues

Demos have high stakes. It is the culmination of an iteration where things happen fast. The product owner and other stakeholders are there to review the features of the product increment and it is the one chance the team has to prove that it has delivered on its commitments. There are logistical challenges, there is a need to make sure whatever is demonstrated actually works, and there should not be any technical issues. Stakeholders are taking the time to review the items that are demonstrated and the project may be reshaped moving forward as a result of this session. It is important to be ready for the demo; and with so many things that can go wrong, there are a few items to pay special attention to:

- Team members should be familiar with how to give the demo
- Team members must be prepared for their part of the demo
- Team members need to be clear as to how to give this demo, who the attendees are, what the purpose of the demo is, and the amount of work involved in preparing and doing it
- Stakeholders need to be prepared to give feedback
- The team (with help of a BA, if applicable) needs to be clear regarding the decisions that are made with the feedback provided

In this context, let's also review a few suggestions for a good demo:

- Ensure that each story is demonstrated by the appropriate team member (the person who led the effort and who is suitable to present)
- Ensure the workload in the demo is properly shared
- Make sure feedback and action items are captured as intended

- Keep the demo short (1–2 hours)
- Demo only items that are working and that were fully tested
- Invest in preparing for the demo
- Focus on the acceptance criteria of each story and its business value
- Work with the end in mind: think about the demo throughout the iteration
- Keep things short and focused, and if possible, interesting and exciting

The Essence of the Demo

Keep in mind that as one of the major steps to enable the inspect and adapt concept, there are essentially three reasons for a demo—to produce answers to these three questions:

1. Is the product that we are demonstrating acceptable?
2. Does the product do what it needs to do?
3. Is there anything that has to change?

Feedback

The demo is at the core of the iteration review and the progress that the team shows in the demo triggers questions, comments, and feedback by the stakeholders. Based on the functionality of each story demonstrated, the product owner leads the way in producing feedback that is meaningful for the team to act on. The feedback is the main driver of the backlog maintenance process that will prioritize items in or out of the main project backlog, determine what will be included in the next iteration, and if required, will change the release plan. All this comes to ensure that moving forward, whatever the team works on fully reflects the progress made, the acceptance of items, the performance of the product, and the most recent and applicable customer needs.

For each story that is demonstrated, the product owner will ultimately make the final decision—which can be one of the following possible options or feedback types:

1. *Accept as is*: The story demonstrated performance as intended and is accepted. The story status is now going to be marked as *complete*.
2. *Accept the story with a minor defect*: This story most likely had large defects that were discovered during the iteration that were not brought forward in the demo. However, stories that suffer from minor defects (whether discovered during the iteration or at the demo) will be accepted and will move to a *complete* status, while a new story is going to be created and put in the backlog to fix the defect. The new story will be placed in the main backlog or in the defect backlog, depending on

the urgency in fixing that defect. For defects that have to be fixed in the next iteration, the story will be brought forward into the main project backlog as part of the backlog maintenance process to ensure that it is prioritized and estimated. Let's keep in mind that since we are talking here about minor defects, it is more likely that the story will be parked in the defect backlog until *if and when* it is applicable.

3. *Accept the story with a change*: The story demonstrated is acceptable and therefore it is marked as *complete*, but a new story will be created to reflect the idea for the change. The new story is going to follow the same rationale as the defect described in the previous item: if it has to be incorporated immediately, the story will be promoted into the main project backlog—otherwise, it will be parked in the shadow backlog until it is applicable to move forward with it.

4. *Accept the story with a new need*: The story is acceptable and therefore is marked as *complete*, but a new story will be created to reflect a new need that was realized. The need may have been realized during the iteration or at the demo and the handling of the new need is identical to the way we handle the new story to reflect change in item 3. The new story will be placed in the shadow backlog for future consideration or into the main project backlog—depending on when it has to be implemented.

5. *Story rejection*: After four different types of acceptance (where the story moves to *complete*), and in some cases a new story is created, this option is about story rejection. A story that fails to deliver the intended (or committed) performance will be rejected by the product owner because it needs major rework. Essentially, the feature was broken or riddled with defects so the story will be rewritten to better represent the customer's needs and it will move into the backlog. Depending on the importance of the functionality that is impacted and the impact of the issue, the product owner will determine whether the rewritten story is promoted into the main project backlog or parked in the defect backlog.

6. *No longer required*: The sixth and last type of feedback is also about a rejection of a story, but this time it is not due to performance issues, but rather because the story that was built and demonstrated is no longer required. Although the iteration's length is only a few weeks long, it is possible that something that was given for the team to do at the start of the iteration may be deemed redundant by the end of the iteration. In such a case where the story is no longer needed, reject the story, move the status to *obsolete*, and if there is a need, add a new story to clean up the code into the backlog. Once again, the product owner will determine whether the new story goes into the main backlog or the shadow backlog.

It is clear that in the process of the demo, feedback, identifying action items, documenting, and creating new stories may lead to loss of information, confusion, and misunderstandings. When team members also act as partial BAs, it is important to define who does what: captures information, follows up with questions, updates the product backlog, and assists in the prioritization and estimating process. In many environments, the outcome of the demo poses a challenge since there is no one person who consolidates the information or who owns the process of capturing feedback. Although the product owner is the person making all the decisions, we cannot expect him or her to capture and document all of the actions required as a result of the demo. In these cases, teams need to come to terms with the fact that it is not embarrassing to have a BA that helps the team handle all of this information.

It is also important to note that every time a new story is being promoted into the upcoming iteration, a story (or more than one) of the same size (story points) has to come out of the next iteration's backlog to ensure that the team's capacity allows them to build the feature. This prioritization process ensures that a trade-off is always made and that when the story is prioritized into an iteration, something has to come out to clear the way for that story. The ultimate decision maker on the feedback and on what happens with a story and when is the product owner.

PRIORITIZATION AND BACKLOG MAINTENANCE

The result of the iteration review and the demo is an updated list of items that need to be performed in the next iteration and in the release. The feedback from the demo moves stories to a *complete* status and potentially produces a set of new stories. Decisions are made about these stories to either park them in the shadow backlog or the defect backlog; or promote them to the project backlog. Stories that are now included in the project backlog will be prioritized (in relation to other stories, to determine the order of the work to be performed) and estimated. This is the point where the team members who will be performing the work of building the story feature will estimate the effort it will take to complete the story's associated tasks. The estimating process ensures that the team knows what it will take to build the story and when they will reach their capacity. Team members should be able to estimate their own work, but a BA can help in providing references, historical information, access to SMEs, and other insights about the estimates. There may also be a need to seek more clarifications about the priority or the estimates from the product owner.

The backlog prioritization mechanism in agile projects goes through stories, features, and other items that have been authorized to be on the backlog by the product owner. The prioritization process is consistent and includes the

Table 7.1 Backlog prioritization considerations

#	Criteria	Who
1	Business value (needs, judgment)	Product owner
2	Technical risk	Developers/Technical team
3	MFS: Minimum feature set for product viability	Product owner and developers
4	Story dependencies	Developers

following considerations: business value, technical considerations, product viability, and other logical dependencies. The BA may become useful in assisting those who engage in providing inputs to these considerations. Table 7.1 shows the four criteria for backlog prioritization. It is important to differentiate the way agile projects prioritize their items from the way traditional, deterministic plans prioritize scope items. In traditional plans, the work is essentially prioritized and sequenced based on the convenience of the development team. Since all work has to be completed by the end of the project, the sponsor does not have any specific preferences about the sequence. Toward the end of the project, the team scrambles to meet the schedule by dropping features. With little to no effort to prioritize features up front, critical features may end up being dropped.

Business Value and Prioritization

The product owner decides the business value and the business value does not change when other considerations are introduced into the mix. What may happen, though, is that the business need may remain high, but other considerations on the list may force the order on the backlog to change. For example, if someone is building a custom home, they can go to the builder and say that since the kitchen is the most important part of the house, they want the kitchen built first—before the basement. The homeowner may even ask the builder to skip the basement altogether; but from a technical point of view, the basement needs to be built first. Therefore, despite the high business priority for the kitchen, other parts of the house may need to be built first.

Business priority is calculated by the product owner and includes multiple factors, such as financial value (e.g., net present value, internal rate of return, payback period) or customer service value. The product owner also looks at the cost of developing/supporting the product, the amount of learning (about the product and the process), and personal skill development. Additional considerations include the amount of risk removed by completing the feature and any legal and compliance requirements. The business priority factors may not be visible to the team members and can be taken into account by the product owner in private. If there is a BA on the team, the product owner may utilize the BA for help in searching for answers for any of the criteria under consideration.

Next, the product owner prioritizes the list of items on the backlog by ranking stories using a value-to-cost ratio, with the highest ones at the top of the list. These considerations will turn the backlog into a risk-adjusted backlog by moving items up and down based on the amount of risk they involve. Figure 7.4 illustrates the logic related to backlog prioritization in relation to risks by measuring items on two dimensions: risk and value. As seen in the figure, when all else is the same, the high-risk, high-value items should be performed first; *low-hanging fruit*—low-risk, high-value items should be done second; followed by the low-risk, low-value items. Finally, the product owner should avoid pursuing the high-risk, low-value items. While this part of the prioritization process often takes place inside the product owner's head, he or she may choose to consult with team members and SMEs for more specific information about some of the items on the product backlog, their risks, the amount of learning involved, and other considerations. This is also the time to consider non-functional examples (e.g., security, graphical user interface design). Once the business priority is in place, technical dependencies are taken into account with more input from SMEs and the technical team. Additional adjustments involve a review of any other logical grouping (such as resource availability, chunking features for expertise, etc.).

It is important to differentiate between items that have known risks associated with them and items that are considered to be risky due to various

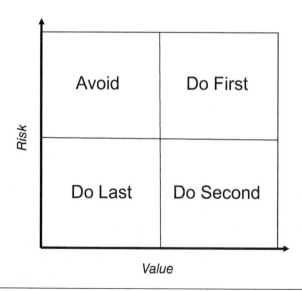

Figure 7.4   Considerations for a risk-adjusted backlog

reasons—such as missing information, not being fully described or well esti-mated, or missing non-functional requirements. In such cases, it may be better off to place these items lower on the backlog until there is more information about them. Another thing to keep an eye on is for changing conditions outside of the project, such as an introduction of a new product by a competitor, or any-thing within a regulatory environment. With team members busy working and estimating, this is another area that can benefit from the extra set of eyes that a BA brings to the table. At the back of the minds of everyone involved in the prioritization process, there should also be considerations that complement customer needs, such as the urgency of getting feedback and the difficulty of the implementation of a user story.

Although it is the product owner who makes the decisions and is the primary contributor to the backlog prioritization process; as seen here, it is not done in a vacuum. At any given stage of the prioritization process, the product owner needs to seek input and feedback from customers, team members, SMEs, and designers in order to maximize the value creation for the customer and to help optimize everyone's workload.

With the backlog being in place, it is important to regularly maintain it to keep pace with the progress, changing needs, performance, and any issues that come up. The product owner reviews the backlog before each iteration planning meeting to ensure that prioritization is correct and that feedback from the last iteration has been considered and incorporated. The process of regularly at-tending to the backlog should be supported by team members and/or the BA to ensure that no considerations have been missed in each round of review. These reviews of the project backlog are the act of backlog maintenance and are often called backlog grooming or backlog refinement.

The initial backlog is created at the release planning stage; at which point it is broken down into releases and iterations. As the backlog gets larger, the product owner and the team group the backlog items into near-term and long-term items. The near-term items (for the upcoming iteration) are going to further refinement in the form of more prioritization and detailed estimat-ing so these items are ready for the work to start. Typically, the design work is done one iteration at a time by applying a fast-tracking technique and by the time the team gets to the start of the next iteration, the design work is com-plete and the team is ready to finalize the tasks and come up with estimates for the upcoming iteration stories. Throughout the fast-tracking process, a BA can become useful in supporting the design work with access to infor-mation from the business, team members, or SMEs. When the team provides the estimate for the upcoming iteration's work, the BA can also support the estimating process and facilitate the matching between the team members and the stories they picked.

Once the items on the backlog are prioritized, they are ready for the team to estimate the work for the upcoming iteration—one iteration at a time. Longer team items can remain at a more vague level though it is good practice to get the team to provide high-level estimates to help prioritize those longer-term items. All involved have to keep in mind that the meaning of high-level or rough estimates is that they provide an idea about the estimates, but these estimates will change once there is more information about these items and the team fully understands and begins work on the longer term items. The priority of backlog items will change as the project moves forward. New information, new needs, new stories, changed needs, stakeholders' views, external conditions, and other considerations are behind the need to keep the backlog a *living document* and keep it current—by adjusting the items on that backlog and their priorities. It is expected that stakeholders will challenge priorities and even change their minds about backlog items, which is part of the dynamic, empirical process to maintain a fresh backlog—reflective of the true, current customer needs—and relevant. The discussions around the challenge of the backlog items and their priorities help get stakeholders' needs and priorities in sync around what matters most and it fosters a culture of group prioritization to ensure that stakeholders share a similar mindset on the project and its product. BAs often play a crucial part in fostering these discussions and ensuring that stakeholders are included in the conversations, that lines of communication are open, and that all views are taken into account. With the lack of a BA, some teams struggle to keep up with all the information and the changing needs.

A BA can also be the balancing act between the product owner and the team. While the product owner is the driver of the priority of work items in the back-log, the team determines its velocity through the backlog. Because of the competing points of view, this can be a shaky relationship since the product owner must remain connected with the team so that he or she does not *push* work to the team. The backlog maintenance process and timing throughout the project have to be set up at the start, along with the creation of the release plan. Backlog maintenance is the key activity that keeps the backlog current. It must include a seamless collaboration between the product owner and the team as well as with other stakeholders. Here too, there are several areas where a BA can add value—including areas related to ensuring the process and procedures to perform the backlog maintenance are followed and areas that enable access to information, stakeholders, and other elements that serve as inputs to the process.

Reprioritization

Reprioritization of the backlog takes place throughout the project. During the iteration, as new information comes to light (i.e., needs, performance), this information will serve as input to the backlog maintenance process and the actual adjustments and reprioritization of items on the backlog happens *between* iterations based on the demo and the review process. To clarify, the word *between* does not imply that there is a break between iterations, but rather that upon finishing the iteration, the demo and review on the last day will determine what will happen and in which order in the next iteration. The need to act quickly, based on demo feedback and changing needs, may introduce the need for a BA to help coordinate the understanding of the reprioritization, changes, and estimates—leading to and through the first day of the new iteration. The reprioritization process follows the same criteria, logic, and considerations of the prioritization process that was described earlier in this section: the product owner is in charge of expressing the business priority; the team members explain, inform, and persuade based on technical considerations and dependencies; and then logical groupings are considered (e.g., cost savings, ease of delivery, lower risk items, etc.). The team needs to demonstrate strong negotiation and communication skills to ensure their views and concerns are heard since it is ultimately up to the product owner to have the final say.

ACCEPTANCE AND DONENESS

Earlier in this chapter, we covered the different *types* of acceptance that stories can have. At the end of the iteration when the functionality of the product increment is demonstrated to the product owner and to other stakeholders, it is critical that they gain an understanding of the extent to which the goals of each functionality have been delivered. For a clear understanding of the product performance, the product owner needs to provide clear guidance as to the acceptance criteria and the level of acceptance (or rejection) of each story demonstrated. In order to enable the common understanding of the needs and the acceptance criteria, agile methodologies put a heavy focus on the *definition of done* to ensure that the team is delivering features that are truly done—from both functionality and quality points of view.

The definition of done is a checklist of valuable activities that are required in order to produce the project's product. It includes a list of activities that add value to the product so it can be demonstrated, verified, and understood at the end of each iteration. A clear definition of done also allows the team to focus on value-added activities that focus on what really needs to be done to build the product. In short, the definition of done is the primary acceptance criteria and reporting mechanism for team members, and it is refined and adjusted based on the reality and project progress. The definition of done spans over every element of the project since there has to be one for the features on the backlog, for each iteration, and for every release. Further, the definition of done ties back to each feature's acceptance criteria, and with that, to the testing and the quality assurance process. The process of defining the acceptance criteria, working with the testers, supporting the testers, and ensuring the demo is successful and that the product increment works involves significant coordination, collaboration, and team cohesiveness. Similar to other elements in agile projects, if team members perform their business analysis tasks to a sufficient extent, it will help the project achieve its goals. Otherwise, a distinct BA role has to be part of the team to support all of the ingredients of ensuring there is a clear definition of done and the acceptance process.

HANDLING CHANGE

Change is good. This is what people say and this is what stakeholders do. Although, in waterfall projects, project managers (PMs) do not like change because it often leads directly or indirectly to increased risk and the derailment of the project. But change is good because it enables the team to get closer to delivering what the customer actually wants and needs. There is no need to reiterate here the importance of change or the ability of agile projects to *handle* change. Further, agile projects do not only handle change, but they are about accommodating changing conditions, adapting to change, and ensuring the delivery of value and success that support the customers' changing needs. The benefits of adapting to change further supports key agile objectives and helps deliver some of agile's key benefits:

- Reducing the risk of building the wrong product
- Reducing the risk of schedule and budget overruns
- Reducing the risk of not meeting the business case or delivering value
- Improving the ability to achieve customer satisfaction
- Improving the ability to learn and adapt
- Improving project efficiency
- Improving project solution reliability and performance
- Improving communications with all stakeholders
- Building more reliable schedules and estimates based on actuals

Even though it is safe to say that agile is (a lot) about adapting to change, agile practices may appear like a conundrum from the outside: if agile is about change, why is it that agile does not allow change during the iteration? The idea behind it is that due to the short iteration length, it is safe to assume that the needs that are realized and expressed at the start of the iteration can serve as a stable, safe, and clear mandate for the team to perform its work during the iteration. While changes can take place during the iteration, it is one of agile's main features that unless the change is a *game-changing* factor, it can wait until the end of the iteration where a process of inspecting and adapting will take place. In short, we *lock in* the scope and mandate for the iteration, and in turn, we will make the necessary adjustments at the end of the iteration. Based on the feedback, progress, performance, and changing needs, the product owner, along with the stakeholders and the team, will shape the way the next iteration (and possibly the release) will look.

How Is Change Taking Place?

The iteration review process includes the demo and the retrospective; and this is where the team gets the feedback on the product increment it builds. This feedback is also the foundation of any adjustments and changes that need to take place for the next iteration's mandate—and potentially the release too. Just because there is a closer working relationship with the sponsor and there is more flexibility throughout the project, this does not translate into a reality that takes the project out of control. Formal change management processes still need to be implemented to ensure that the feedback is considered and that the mandate to move forward represents the most relevant and recent set of customer needs. For these reasons, formal sign off on adjusted features, schedules, budgets, and resource allocations must still take place. The mechanisms to allow the feedback and changes to be incorporated into the plan to move forward must take place in a fast, nimble, and consolidated way—and all of that must not compromise the quality of this process. To handle the changes that were introduced throughout and at the end of the iteration, the process involves negotiating all of these changes at the end of the iteration as a group—putting them together into a single, omnibus change authorization form. Changes that are proposed for a future iteration will be *parked* in the shadow backlog. Similarly, defects that are raised to be fixed in the future will *park* in the defect backlog. However, any changes or additions to the upcoming iteration will need to be approved, considered, and estimated. The change authorization and review process must be fast so that it is timely and enables the team to finalize the estimates for the upcoming iteration. This process should also engage the stakeholders who need to be engaged (and who are affected by the change) about the proposed changes to get their points of view, buy-in, and signatures. The signed change authorization provides the mandate for the new backlog and updated release plan and the upcoming iteration plan.

With the review, the stakeholder engagement process, and the need to estimate the work and impact of the approved changes, the change process in agile projects is similar in nature to that in a waterfall environment. Key differences are that in agile projects the change process is done in an organized and periodic manner at the end of the iteration unlike the somewhat random and disorganized form it has in waterfall projects. In both environments, however, there is still a need to go through the process and ensure that the result of the change process is effective, realistic, and clear. Also, in both cases there is a need to get all necessary and involved stakeholders' signatures. Another difference from waterfall is that in agile projects, the change process is more consolidated—it keeps the paperwork to a minimum.

The change process in agile projects is simple to perform in theory; however, there is a difference between theory and reality where there are two factors that make the change process more challenging:

1. *The emotional factors*: Stakeholders get *attached* to needs and features and, at times, it is hard to *sell* them on the idea of a change—especially if the change is about to remove or delay the building of a feature that they asked for or that they are associated with.

2. *The fast pace of this process*: It is not that waterfall projects allow ample time to process change requests, but the change process in agile projects takes place under almost extreme time constraints since the change needs to be processed in full from potentially as late as after the end of the demo—and it must be completed by the next working day when the next iteration begins. The rushed process introduces more pressure because not all stakeholders will be available, engaged, or supportive of the proposed changes.

To support the change process so that it is done properly and produces the timely and desired results, the BA can assist with the *traditional* aspects of the role, including reaching out to stakeholders, streamlining communication, and generally expediting the process.

RETROSPECTIVES: APPLYING AND FINE-TUNING LESSONS

One of the final acts of the iteration is the retrospective. It takes place after the demo and it is a process review or a lessons-learned session. Similar to the lessons-learned process in waterfall environments, the team reviews what worked well, what did not work well, and whether anything should be done differently in the next iteration. Feedback from team members is collected, the information is consolidated, an agreement (consensus) is formed, and action items are articulated to take place in the next iteration. Generally, it is a good practice to come up with a couple, but not more than a handful, of items to change and apply in the next iteration in order to make sure that there is control over the continuous improvement process, that the team can focus on the areas for improvement, and that there is the ability to check how effectively the new lessons are implemented. The retrospective is about looking back at how the team conducted itself and looking forward at how the team can get better and continuously improve its processes, communication, and cohesiveness.

It can be beneficial to hold two retrospective sessions right after the demo. The first part should include team members only—where *raw* findings and discussions take place, thoughts can be formalized, ideas organized, and recommendations consolidated and put together. The second part should include the product owner, who will help convert the ideas into actions for the upcoming iteration and provide the mandate for the improvement initiatives. The main rationale for *breaking* the retrospective into two sessions is to allow for a free flow of information within the team in the first part and to show respect to the product owner's time by including him or her in the second part. The recommendations for actions should be clear, small, and simple enough so that they can be implemented and measured throughout the upcoming iteration. The next iteration's retrospective will measure the application and performance of the improvement action items to ensure they were implemented properly and that any side effects were addressed and minimized. The BA can provide input to the retrospective based on observations he or she made during the iteration regarding the team's performance or anything else the BA deems to be relevant. The BA is also *ideal* to review the application of lessons from the previous iteration into the next and provide feedback about it. If there is no dedicated BA role within the team, it is important to ensure there is a clear understanding as to who should collect information, make observations, and track the application of the previous iteration's lessons.

The retrospective is the time to reflect on the results of the process and the team's performance during the iteration. The tone and recommendations of the session will also be impacted by the product's performance. This is one of the rationales behind having the retrospective after the demo. The retrospective should add up to a total of about one to one and a half hours with a clear and simple agenda:

- A review about whether the team met the iteration's goals, to what extent, and how
- Collect performance metrics, including team velocity
- Review quality of estimates, plans, design, and other elements of the team's work
- Review the application of improvement stories from the previous iteration
- Analyze current processes
- Find a couple of things to improve in the next iteration, along with who *owns* these items, if they are relevant, and how to measure their success
- Review of conduct, communication, and process: what works well, what does not work well, and what the team can do better next time

- Get the product owner to provide feedback on the iteration, present findings to the product owner, and get a mandate for the action items

The retrospective is also a good time to start the process of getting ready for the next iteration, including preparing to complete necessary documentation and ensuring that the information about the next iteration is clear, such as understanding constraints, velocity, resources, changes, features not accepted, new stories, and product questions, hurdles, and issues.

The retrospective is a very important meeting; it serves as a *tune up* and a realignment session for the team. It helps the team focus on improvement ideas, clears the air about issues, and reflects on how they can perform better moving forward. While the retrospective can help achieve economies of scale and improve team velocity, this is not the main goal of this session. It is in place to ensure that issues, problems, friction, misunderstandings, and challenges do not weigh down the team and that the team can clear these things out of the way so that the team members can optimize their performance.

RECAP

Every agile work product and ceremony is in place for a reason and can be performed sufficiently with or without a BA. This chapter has pointed out most of the key events in an agile project and showed different ways to apply business analysis skills or to have a BA contributing to these events. The reason that the daily Scrum was not included here is because there is no need for any specific actions by the BA during the meeting. After it is over, however, those who remain there after the meeting to discuss hurdles and impediments can use the help of a BA in addressing any problems, looking for answers, and accessing the relevant SMEs or stakeholders if needed. However, throughout agile project events, if the team has a BA, there is a need to establish clear rules of engagement between the BA and the PM (or the Scrum Master). Whenever the BA gets involved in a process—especially when it comes to facilitating or acting in an impartial way—it is important to clearly define the roles of the BA and the PM. Further, at times, it may be easier for the PM to be an impartial contributor to a process, while the BA is more engaged in product and requirements-related considerations.

8

TESTING AND SOLUTION
EVALUATION

Agile testing, on the one hand, is quite straightforward: there is a need to evaluate the solution and there are team members who do it. On the other hand, there is a lot of confusion surrounding testing in agile projects: who exactly needs to do this work, what does the business analyst (BA) do to help the testing process, and above all—how to handle the growing volume of work and the increasing role of the testing team in project success. This chapter provides answers about the changes we should expect with testing activities and the role of the testers in agile projects, and it looks for ways where the BA can add value to the testing process.

It is important to note here that although agile projects can be performed in virtually any type of industry and environment and can be used to build essentially any type of product—the context in this chapter leans toward technology and software development projects. In addition, although testing needs to take place in all types of projects (agile or others) and in all indus-tries, the illustrations and examples in this chapter focus on testing activities primarily in software development projects.

We will start this chapter by reviewing the agile approach to quality, key con-cepts of agile testing (including the testers' earlier and proactive involvement in the project work), the definition of *done* (DoD), the increasing work volume of the testers in agile projects (especially due to growing regression testing), and the need for automated testing. The areas listed here all fall under the scope of work for the testers or the quality assurance (QA) analysts; and as we progress through the chapter, we will show how the BA can help the testers and support them in producing value for the project.

AGILE APPROACH TO QUALITY

One of the main areas agile focuses on is quality: it is a value-driven approach that delivers value to the customer throughout the project. During each iteration we produce a product increment that is a production-quality slice of the product. Each increment represents value to the client and it is shippable—in the sense that, should the client choose to use it, they can. The iterations are *bundled* together into product releases that are set to take place in intervals that maximize the value to the client. As product increments are released, they can be used by the client in the interim, helping the client realize value early and reduce the overall cost of the project. The cyclical nature of *build-test-review-release* allows the client to get a better picture of the actual progress and ensure quality is delivered. Beyond the early realization of value by the client, the iterative and incremental approach adapts to change, reduces risk (of building the wrong thing or building it wrong), and allows the client to further focus on managing value by allowing an efficient reduction of scope if needed.

The quality focus takes place through test-driven development where the testers are part of the development team; they get involved proactively, early, and during the course of the project. Testing takes place throughout the project, full regression testing takes place within each iteration, and there are frequent customer reviews and feedback cycles to help ensure that the right product is built.

Test-Driven Development (TDD)

To further focus on quality, many agile teams follow TDD principles where there is a heavier focus on the testing process since it essentially reverses the way we are used to doing things. While with most project teams, the design work leads to a development effort through which the testers write test cases in an attempt to ensure that the intent of each feature is delivered, TDD focuses on testing. It calls for an increased focus on defining the acceptance criteria for each feature—and these acceptance criteria serve as input to the design, development, and testing processes. In TDD, the development team writes feature verification test cases before starting to build the features (known as the *test first approach*) and the test cases become part of the solution design.

There are a few conditions that must be in place for TDD to be executed successfully:

- *The customer must have a clear vision of what they want:* Without a clear vision, it will be hard to define *acceptance criteria* and the acceptance criteria that is defined may not represent the true needs of the customer. Without clear acceptance criteria, it will be impossible to build (any or effective) test cases.

- *Developers, designers, and testers need to have strong people skills*: Every agile team needs to have technical team members with strong people skills and the ability to work with the business and the client; but in TDD, this need is accentuated due to the key need to have the acceptance criteria clearly elicited and translated to test cases. While a BA can provide valuable help to the process of engaging the customer, that may not be enough. Relying heavily on the BA may lead to a dramatic increase in the BA's workload because the BA will become a translator that needs to pretty much own the process of writing the test cases. It will also create a heavy reliance on the BA, slowing the process down, and potentially creating a bottleneck. In turn, this may tie up the BA from helping elsewhere in the project, causing further waste and additional challenges.
- *Automated testing and reporting tools for at least partial test coverage*: This chapter discusses the need for testing automation in agile projects, but in TDD the need for testing automation is even more critical. The team will not be able to handle the need for full regression testing in each iteration without automating at least part of the test cases. The BA can help the team determine which tools to select and how to maximize the value that the tools produce, along with what level of test coverage and which test cases require automation.
- *A stable testing environment*: The lack of a stable environment will undo most of the benefits that TDD offers.

THE GROWING AND EXTENDED ROLE OF TESTING IN AGILE ENVIRONMENTS

Agile teams—and specifically Scrum—have two leading sets of important characteristics: (1) they are technically excellent and cross-functional so all skills to perform the job are within the team; and (2) they are self-directed and self-organizing so they need to find ways to get the job done. When referring to the team, we mean the entire team—designers, developers, testers, and all other roles—without a distinction of use against them or of developers against testers. With that said, the role of the testers (also commonly known as the QA analysts)[1] is significantly enhanced from the QA analyst role in waterfall projects; and in agile environments, the QA analyst role goes well beyond just writing test cases and finding defects in the product.

The role difference for testing in agile projects starts with two sets of considerations: (1) the expectation that we will build product features in a specific order; and (2) the notion that the testing professionals need to effectively engage during the iteration even before anything has been built. Let's start with the order or sequencing. In waterfall projects, the testing team reviews the requirements and the product that is being developed and then they write test cases that are subsequently executed. The testing process takes place late in the project after the development team has completed the building process (at least for that phase), and testing commences to look for defects. The order that we test the product does not matter much so features are being tested in whatever logical order the testing team comes up with—at times with little regard to the importance, complexity, or other specific considerations related to the features being tested. From the client's perspective, there is also no difference as to the order by which features are being tested since the entire product is going to be released at the end in the form of a *big bang*.

The second consideration is that testing in agile environments is much more proactive than in *traditional* projects. Many environments engage a TDD approach that is so proactive and testing-focused that it essentially reverses the order of doing things; it starts by clearly defining the acceptance criteria. At this point, the testers write test cases and then the developers build a product with features that specifically aim for the test cases. This increases the chance that the team will build a product that truly addresses the needs of the business. Even on non-TDD environments, agile teams are more focused on an earlier, increased, and proactive approach to testing.

Proactive Approach

In *traditional* environments, the testers only get involved late in the project— once the entire building process of the product is complete. A simplified view of the testers involvement includes the following steps: the testing team gets to review the requirements document and the completed code; they write test cases that need to reflect business needs; and they execute the test cases (i.e., perform testing) to verify that the product does what the requirements documentation says it should do. At this point, the testing team will report on the product's performance and the defects they find.

This process has a couple of inherent flaws that set it to systematically fall short of expectations:

1. In many environments, the QA team does not come in direct contact with the customer or the business stakeholders and, therefore, they write the test cases based on *second-hand* information that comes from the requirements documentation. Although it is expected that the requirements documentation was done properly and truly reflects business needs, it is likely that the testers will miss nuances and underlying needs that could have been picked up had they had direct contact with the stakeholders behind the requirements.

2. Once the development work is complete, the testing team gets the first chance to see the product they are going to test. In a way, this is the

definition of *reactive*, as whatever the testers find is already an after-thought—after the fact. Winston Royce, the *inventor* of waterfall, said: "I believe in this concept, but the implementation described above is risky and invites failure."[2] He articulated an approach to software development, but he also realized the inherent problems that process might lead to. He specifically referred to the testing process: "The testing phase which occurs at the end of the development cycle is the first event for which timing, storage, input/output transfers, etc., are experienced as distinguished from analyzed"[3]—pointing out the flaws in the proposed reactive process. Royce also noted that the design iterations are not confined to the successive steps and, therefore, we should expect that by the time we get to test the product in a waterfall project, it is simply too late. Figure 8.1 illustrates the flaws that are inherent in waterfall projects where the findings from testing circle back to the earlier design stages.

Working Together

Agile testers have a variety of responsibilities. The testers get involved at the very beginning of the project and they work closely with the BA and with the developers. With the team being cross-functional, it is cohesive and works closely

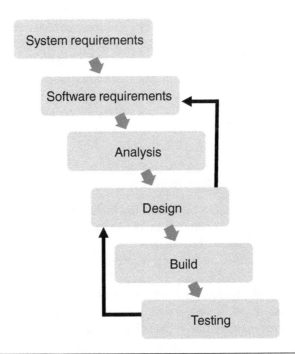

Figure 8.1   Testing challenges in waterfall

together. Beyond the core role of the testers to build test cases, they also help the product owner ensure that there is clear acceptance criteria for every story and feature on the backlog—and they do it by challenging assumptions and asking questions so that the business requirements are clear enough for the acceptance criteria to be articulated.

Early Involvement, Planning, and Estimating

As part of working cohesively with the rest of the agile project team, the testers participate in the planning and estimating process on both the release planning and the iteration planning levels. In addition to estimating the complexity and work effort from a testing point of view, the testers identify where there is a need for more complex and negative test-case scenarios. Identifying complexity is an area where testers can add value to the development team since testers tend to look for ways to *break the product*, or at least to ensure that it works well, while developers are naturally drawn with their focus to the *happy path* for each story and requirement since their efforts revolve around ensuring that what they build works as required.

The early involvement of testers also includes helping the product owner articulate acceptance criteria for user stories and, subsequently, the testers' role continues throughout the project by providing context and advice to the team and the product owner on changing or enhancing user stories so that they continue to reflect the business requirements and the changing needs. The testers work together with the developers, helping as needed with unit testing and ensuring fast feedback rolls back to the team and stakeholders as a result of the rest of the testing effort. Through the process, while working on stories and features, developers consult the testers about acceptance criteria or about the expected behavior and functionality from the user's perspective. This activity makes the development process more focused and leads to a reduction in the number of defects.

Increasing Role of the Testers

In *traditional* projects, testing takes place late in the process—after the product is built. The verification is the comparison of software to its technical specification, typically performed by a systems analyst, and the validation process focuses on comparing the software to the business requirements typically led by the BA. The Traditional V-Model,[4] as shown in Figure 8.2, illustrates the waterfall testing process. On the left side, we can see the up-front and comprehensive business and technical document, while on the right we can see the comprehensive levels of testing that mirror the activities on the left. In a V-Model situation, the developer

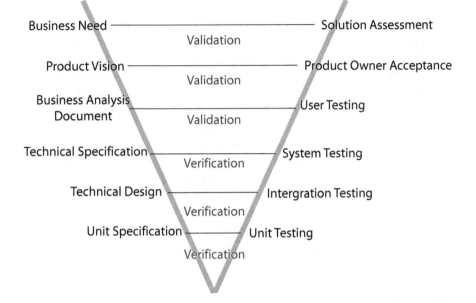

Figure 8.2   Traditional V-model

submits *working software* to testing where the tester finds defects and integration issues (one functionality does not work with another feature). At this point, the tester will fix the defect and the integration issue.

When it comes to the testers in agile environments, in addition to being involved earlier and on an ongoing basis, the amount of work they have grows from one iteration to another and from one release to another. Agile testing involves unit testing, which is performed by the development team during the iteration and involves an automatic test on code level that (ideally) runs every night after new code is added to check that no bugs have been introduced throughout each day's development effort. Functional acceptance testing (manual or automated) takes place by the testers during the QA portion within the iteration. Performance testing takes place at planned milestones and after releases, and system validation tests are done after releases and during deployment. The agile testing process is about the developers submitting *done* software increments to testing and the testers validating whether or not it works.

Regression Testing

One of the largest *contributors* to the increasing role of the testers in agile environments is regression testing. Regression testing is an old practice in software testing (though its use dramatically increases in agile projects) that comes to ensure that whatever was previously built in a past release is not going to be broken because of new changes that have been built into the product. Many development teams (especially, but not exclusively, in waterfall environments) do not carry out regression testing, mainly due to a lack of time or because the core functional testing was itself pushed to the final stages of the release cycle. With that said, there is tremendous value in performing regression testing in reducing challenges at release time to a minimum.

In most environments, regression test libraries can be built from existing test cases that were developed throughout the project. Functional, unit, integration, and build verification tests can all be incorporated into regression tests; but it is quite obvious that over the course of a complex development project, the number of these test cases grows dramatically, making it next to impossible to perform the testing, especially in later releases in the project. It is therefore critical to incorporate automated testing tools for full-scale regression tests (discussed later in this chapter).

What Is *Done*?

In agile development environments, we aim to deliver *potentially shippable code* that represents value at the end of each iteration with the goal of closing each iteration with no leftovers and no dependencies—where everything has to be *done*. In waterfall environments, done is a fairly simple achievement: we follow a series of tasks and we call it done when all of the steps are complete; at which point, we release the product. In an agile environment, we follow these steps for each iteration—and even daily—hence the need to make sure that what constitutes value and the DoD is clear to everyone.

Testers have an important role in helping the team and stakeholders articulate what is done for the iteration, the release, and the project as a whole. For the iteration, done means that the product increment has been tested and passed acceptance testing or client approval, it passed internal iteration review, and it is of a *shippable* quality—meaning that the client can use the product now should they choose to. For the release or the project, done means that the iteration, reviews, fixes, and acceptance are all in place and that the users or client can utilize the deliverables as intended to the right extent.

From the team's perspective, done means that all tasks are complete; all tests have been run and passed; walkthroughs are complete on all fully supported products; performance test results have been reviewed, and defects are closed by testers; user experience (UX) review is complete; and the product is accepted by the product owner.

The DoD starts as a simple list of team-defined completion criteria that must be performed before a user story can be considered done. While it's not the direct responsibility of the testers to define done, the testers often monitor the work that is performed by the team in order to ensure that each completed story meets the criteria for done. A supporting practice includes reviewing the team's DoD before the start of work on every new user story in order to clarify expectations. The DoD can potentially *evolve* over time, along with changing project needs.

The responsibility on defining what constitutes done does not necessarily lie with any one team member. We saw here that the testers are natural candidates to facilitate this process; however, the BA can also help in this task.

Automated and Exploratory Testing

Regression tests are part of a growing test suite that needs to be run regularly to monitor existing functionality. Therefore, the growth of regression tests has warranted improvements in test automation. The return on investment of automation is very high for regression tests, but this does not mean that all tests have to be automated. There is still room for combining some manual testing into the mix, especially for less used functionality, less important features, and test cases that are less likely to be used in the future. It makes no sense to automate 100% of the test cases.

Since many test teams embrace exploratory testing, it helps to create new tests for defects that are found through the exploratory testing and then add them to the regression test. This will help ensure that these defects are not missed in subsequent test passes. According to the Agile Alliance, exploratory testing[5] is an approach that is characterized by emphasizing the tester's autonomy, skill, and creativity. It recommends performing test-related activities (e.g., test design, test execution, and interpretation of results) in an interwoven manner throughout the project, rather than in a fixed sequence and at a particular phase. It also emphasizes the mutually supportive nature of these techniques and the need for a plurality of testing approaches rather than a formal *test plan*. When defects are resolved, the tester verifies not only the core fix, but also surrounding areas that might be impacted by the fix—which is why the effort is called defect regression.

Many practitioners say that automation is *the tester's best friend* because it provides consistency, repeatability, scalability, and better test coverage of the product's functionality. These benefits become even more pronounced in projects with shorter iteration cycles where the testers have little time to perform both full functionality testing of the new features, as well as full regression testing for all previously implemented product increments. As full regression testing work grows significantly with each passing iteration, any automation that is applied to these tests would significantly reduce the pressure on the testers, as well as the rest of the team. Testing automation can also help provide rapid feedback. Without automation, manual performance of these tests becomes a very monotonous and error-prone task. Automation can help the team deliver testing work both efficiently and effectively.

Limitations with Testing Automation

We should not take an *autopilot* approach to testing automation. First, we should not automate all test cases because it will be inefficient to automate cases that are rarely used and those that have low impact if they fail. It is recommended to use the 80:20 rule—automate the 20% of cases that exercise common functions that are used most (80%) of the time. This efficient approach to testing automation helps ensure that the core solution works in the least costly manner. If there is a need, we can then supplement the automated testing with manual testing on the uncommon test cases.

The way to determine which and how many test cases to automate is by preparing a risk profile of the solution (with the help and insight of the BA) and identify areas that represent a higher level of risks, including:

- Areas where new technology is implemented
- Areas with major changes to existing shared components that may impact other products or systems
- Custom development work
- New interface types or experiences that have not been used by the customers before

We should then try to identify areas with a low risk profile, such as:

- Areas using existing shared components that are stable
- Implementation of packaged software
- Minor changes to existing components or designs

Naturally, we should focus our testing efforts on the higher risk items, but we should not completely ignore the lower risk items, applying the 80:20 rule. There are some circumstances where we should look into automating more test cases; for example, when the solution is expected to endure through many future releases (and we have the capacity to automate more test cases). In this case it makes sense to go through the trouble of automating more of the less common test cases, but we must keep in mind that even in this situation it will be a waste of resources to test everything.

An *autopilot* approach can backfire by not catching all of the defects. When we run the same suite of tests repeatedly at the end of every day, the testing process may turn static and over time developers may inadvertently learn how to *pass* a fixed test library—and by that, compromise the benefits of the regression testing. Just as manually testing through the entire application will become progressively impossible, over-automating testing may make the process too predictable, overly focused on commonly run test cases, monotonous, ineffective in finding defects, and expensive. A middle ground is the right approach by complementing the automation with some good old-fashioned manual tests.

Additional Challenges with Testing Automation

- The fact that tests are automated does not mean that all tests are reliable (as they are often not sufficiently planned due to time constraints). Generally, it takes longer to fix unreliable tests and these tests may lead to multiple issues downstream as a result of false positives or by the tests *finding* defects that do not exist.
- Automated user interface/UX tests are harder and more expensive to build, slow to execute, and more complex to maintain.
- It is strongly preferred to have the automated tests run automatically (through continuous integration) rather than manually, as in the latter they may not run regularly—leading to failed tests.
- Productivity reports that are produced by automated testing tools may be misleading, forcing us to focus on the wrong areas (e.g., measuring the number of test cases executed per day can encourage us to invest in running many tests that add no value).
- Due to the extra work and costs associated with building and maintaining automated tests, they do not always yield the desired efficiencies.
- Team members who do the automation must act as effective consultants: being collaborative, accessible, and resourceful, in addition to having the expertise to put together and produce effective automation solutions. Due to resource and time constraints, these team members may need the help of the BA in complementing their skills and overcoming the constraints they are facing.

- There is such a thing as *too good* in testing automation: it may be so successful that it points out all of the meaningless problems to address (noise), leading the team to work on less important issues. It will end up costing if the team's focus shifts to things that are minor by nature. It is also an indication of efficiency problems with the testing automation since it is performing in *overdrive*, equivalent to gold-plating.

Moving Forward

Regression testing is part of a comprehensive testing methodology that has to be efficient, cost-effective, and dynamic. The testing methodology is driven by a clear testing strategy that is driven by risk prioritization and a strong focus on up-front planning. The methodology should incorporate a variety of testing, including partial automated regression testing, well-designed front-end user interface and UX automated tests, and targeted unit testing. Such methodology can help reduce the chance that parts of the product will go unchecked and, in turn, it can help reduce the chance for budget overruns by keeping the team on track and reducing significant defects from making it through the testing process and hurting our product, performance, and bottom line.

HOW TO DEAL WITH DEFECTS

Unlike what some people believe due to misconceptions—agile projects will have defects. It is less likely that major defects will linger beyond an iteration to impact key functionality, but there are defects in agile projects. The reason it is less likely that severe defects will occur is due to the tight controls that are built into the agile processes and ceremonies. The planning cycles complement each other—down from the high-level release planning; through the realignment, reprioritization, and validation of the high-level planning with detailed iteration planning; and ending with the individual team members' daily practices. The next set of controls comes from the daily routines, the help that team members provide to each other for any types of impediments, the daily update to the project manager (PM), and the daily build and testing where possible. Progress is then updated and, with the help of the BA, new and changing needs are prioritized and planned into subsequent iterations. Without making team members feel *oppressed* and micromanaged, agile offers a very strong ability to understand the current status, realize the impact of events, and control the direction of the project.

Defects can be found during several stages and should be handled differently. Iteration-specific defects should be fixed immediately within the iteration so that it is possible to continue on with the current iteration. In this case, the team will reach the end of the iteration with a working set of functionality that is ready to demonstrate and will represent business value. With all these processes, however, there is a chance that due to various reasons—including missing something through the planning process, performance issues, or changing conditions—defects that are created during the iteration may not be resolved before the iteration-end and will be left over at that time. For defects that are left over at the end of an iteration, the team (and sometimes the BA) will enter the defect into the defect backlog, and transfer the defects to the project backlog in groups (bundles) in order to schedule them like they would new features (and stories). These bundles of defects will be grouped together (again, with help from the BA) in any type of logical grouping to ensure that they are introduced into a future iteration based on their priority, technical area, relevance to the theme of the iteration, and any other considerations related to the team, dependencies, or organizational needs.

Are These Always Defects?

Sometimes it is hard to tell when a defect is a defect; it is possible that the team built something the wrong way or it may be that the team built the wrong thing. However, it could be that it is not a defect, but rather a new need or a new feature that the customer wants. The BA can help the team communicate with the appropriate technical subject matter experts and stakeholders to minimize these types of misunderstandings—but on a proactive basis in real time. The involvement of the BA could be to seek ways to provide clarifications, walkthroughs, reviews, and proper documentation of the stories and requirements. By saying proper, it does not imply that there is a need to be heavy-handed with documentation, but rather it comes in the context of providing sufficient buy-in, participation, and clarification with all those involved in working on the requirements so it truly reflects business value.

Despite all the mechanisms, we need to keep in mind that in agile projects the scope is not clearly defined in detail up front; and for that reason, agile projects are prone to feature creep, especially when the scope is not managed well. The fact that in agile projects the details of the scope are not finalized at an early stage and that we allow ambiguity regarding these details until further information is required, does not imply any type of looseness on the scope or the objectives. We should still stand on guard against scope creep and ensure that features do not make it into the project scope as we go without being accounted for—just because stakeholders realize things along the way. Unaccounted for functionality (or scope creep) may veer the project away from delivering value to the customer.

With that said, sometimes we just need to accept *clarifications* in the name of customer satisfaction even if they start to feel a bit like new features. These clarifications should not be injected into the scope systematically, but at times, we might turn a blind eye and allow these features into the scope as long as they are in line with the mandate and they increase the value delivered to the customer.

Managing the area of clarifications, new features, and defects through the product backlog and reprioritization process can help us gain control over the issues that might be triggered by them. Proper backlog management, prioritization, and reprioritization are areas that can be significantly enhanced with a BA in the project. Best practices in the area of procurement management can also support the efforts to contain the issues that may be triggered by scope creep by having a fixed and variable scope or a time and materials contract to allow accountability for both the portion of the project representing value, as well as any additional scope items.

A Reminder: Testers Are Part of the Development Team

Agile teams are multidisciplinary—and this includes testers. It is not the fault of waterfall, but the way things get done in waterfall environments that encourages a division of roles within the team and practically splitting the project team into sub-teams—including a development team and a testing team. This type of structure does not encourage collaboration and it drives the culture of *us versus them* that is prevalent in many environments. When the team is broken down by specialty, it compromises communication and reduces the team members' ability to work together effectively. The breakdown causes a delay in the testers' involvement in the process and hurts the team's ability to properly capture customer's needs and define acceptance criteria. The result is less context and understanding of the true needs by the testers and difficulties in putting together the right test cases. Later in the process, the disconnect between developers and testers will lead to more misalignment in the testing process. On the project level, these impacts can potentially lead to more defects found, more defects missed, a risk of building the wrong thing, as well as cost and schedule overruns.

Having a cohesive team that includes the testers will result in the following benefits:

- It will set the stage to promote a better understanding of the solution overall
- It will mean less reliance on accurate documentation that has to be produced in the early stages of the project so that the testers know what test cases to build; the testers' reliance on documentation also means that the information that they get is not firsthand—increasing the chance of omissions and inaccuracies
- It will offer early involvement by the testers with direct access to the business stakeholders, which can help improve the overall knowledge of the risk profile of the solution
- It gives a proactive and collaborative approach that can provide more insight for the testers into solution areas that were troublesome in development and with it more effective application of their efforts
- With context and understanding of business needs (i.e., access to acceptance criteria directly from the customer), design, and development work, there will be fewer delays in getting effective testing started
- From a technical point of view, testers can participate in early design reviews, inspections, and other activities—resulting in a better ability to catch defects earlier in the repair cycle

The benefits of building a cohesive team that is multidisciplinary and that includes the testers are clear—and these benefits can make the difference between success and failure. Despite the attempts in agile environments to build such teams and to include the testers early and on a proactive basis, it does not always work as planned. Organizational influences, resource allocation issues, and cultural issues that contribute to the divide between developers and testers may interfere with the attempt to build those teams. The presence of a BA in the team is not likely to overcome all these issues, but it can help with setting expectations, helping identify roles and responsibilities, improving communication with the business stakeholders, and bridging some of the gaps between the members of the team.

MORE ABOUT AGILE TESTING

- *Agile testing offers a few key advantages*: The responsibility for quality lies with the entire team; the entire process, including the testing activities, support and accommodate changing conditions; and feedback cycles are short and similar to other parts of the agile project—it is lean on documentation.

- *Continuous delivery (CD)*: This is a set of practices, processes, and tools that help build software applications that are available for deployment or release at any point in the process.
- *Continuous integration (CI)*: This is an important part of CD and involves producing a clean build of the system—potentially several times per day. The test strategy and its planning have to be adaptable and flexible to new modifications that are incorporated after each iteration. The role of the tester has to be more flexible—becoming more of a quality champion, rather than a problem finder. In CI environments specifically, and generally in all agile projects, implementation of automated testing can significantly reduce the unnecessary repetition of manual tasks.
- *Participate in demos and reviews*: It goes without saying that the testers actively participate in both release readiness and demos (at the end of iterations) in order to help the team demonstrate the user stories that were completed during the iteration.
- *Support the developers*: With developers focusing on developing their assigned user stories and fixing defects, testers should write test cases, clarify questions from the product owners, and automate the previous iteration's sprint stories. Since developers typically have less time to explore the complete functionality of their user stories on their own, testers help them better understand the user stories because testers are more aware of the complete functionality and the acceptance criteria for each requirement. This collaboration is often supported by the BA and it helps free up the developers to handle technical questions that emerge.
- *Analyze user requirements*: Testers can help the team better understand user requirements since it is common that part of the testing process requires them to use the application in a similar fashion to the way end users would.

Testing and the BA

The testers should prepare a high-level test strategy and plan to guide the team through the project—and with no specific guidance as to who *owns* this process, this work can be supported by a BA. This is only one of many potential areas where the roles of the testers and the BAs may converge. The role of the testers and the QA role can also potentially overlap with the BA's role since those two roles share many responsibilities, skills, and objectives. A simplified view of the shared areas, touchpoints, convergence areas, and overlaps between the roles of the BA and the tester in an agile environment is presented in Table 8.1.

Moving forward through this chapter, we raise some challenges around agile testing that can be (at least partially) addressed by support from a BA and we further focus on the touchpoints and the relationship between the BA and the agile test team.

Table 8.1   Comparison of QA and BA roles

Business Analyst	Tester/QA
Analyze user stories from a business point of view	Test for completed functionality each iteration from the end user's perspective
Facilitate the creation of project and iteration backlog	Define and refine acceptance criteria for each user story
Facilitate the backlog prioritization process	Regression testing: ensure previously completed functionality has not been damaged by the new product increment
Help articulate new and changing needs and facilitate the planning and prioritization process of new items	Ensure complexity and acceptance criteria are clear for any new need and story
BA role is integrated with testers' role, other team roles (developers), the Scrum Master/PM, and the product owner	Testers work closely with the product owner to clarify anything related to acceptance criteria with developers and the BA

The areas that overlap may introduce common issues that are associated with the touchpoints: from misunderstandings, to a lack of clarity about specific roles and responsibilities, to other team members failing to give BAs and/or testers a sufficient amount of attention—especially in large teams—with many changing needs and heavy workloads.

Just as people who have two roles tend to be drawn to their primary role, there are risks associated with anything that is being looked after by more than one role or individual.

CHALLENGES WITH AGILE TESTING

Before discussing some of the main challenges with agile testing and the ways to deal with them, let's take a quick look at the main differences between agile and waterfall projects and their direct and indirect impacts on testing. Table 8.2 provides some context about the characteristics of and impacts on agile testing, including a look at many of the challenges in the testing process in agile environments.

Table 8.2   Agile versus waterfall highlights

Considerations	Waterfall Characteristics	Agile Characteristics	Impact on Testing in Agile Projects
Development cycle	Long, since the requirements are defined in detail up front.	Short and time-boxed cycles (known as sprints or iterations).	More frequent testing makes it easier to find defects and to find them earlier.
Scope and requirements	Big set of requirements finalized early in the process for the entire project with the expectation that the customer knows what they want and need and that the performing team got it as intended.	Limited scope per cycle; in each cycle we produce a working slice of the scope, called product increment.	Easier to test and to find problems, but more frequent cycles of testing.
Controls and proactiveness	We wait until the software is fully developed before performing testing, which could be several months after the start of the development process (and after the *finalization* of requirements).	Many environments do continuous integration: code is checked in daily, or even several times per day, and re-compiled at the same time.	Frequent (essentially ongoing) testing requires more capacity (and automation). Without the added capacity, the testing process itself may become a bottleneck. Since the product is constantly changing, there is a more frequent need for testing, including regression testing for every iteration.
Change	Change control processes are in place to assess the change impact reactively while things are in motion. Generally, there is an immediate incorporation of changes with little regard to their relative priority. Testing often suffers since there is limited ability to build the correct test cases.	Inspect and adapt: changes (i.e., defects, new needs, changing needs) are considered as they are introduced and prioritized into the backlog *in between* iterations as we complete one iteration and before the start of the next one; this maximizes the value we produce toward the latest customer needs.	Indirect impact on testing, as it becomes easier to identify change impact and to test the change component.

Bottom line	Waterfall projects are very structured (at least in their processes) with an attempt to fully define requirements up front, supported by a reactive, and at times, slow and cumbersome change control process.	Set up to maximize the benefits of the product we build and to ensure that the customer gets what they really need, even if it is slightly different than what was initially thought and defined.	

Agile Testing Challenges

The smaller development cycles and the frequent testing provide a clear set of benefits for the agile project, but it comes with a price: an increased need for testers, testing environments, reliable preproduction environments, and testing automation tools. Let's review a list of some common challenges that agile testing teams face, along with ideas on how to mitigate these challenges. Some of the mitigation ideas involve help and support from the BA.

Frequent Builds

The high frequency of builds may lead to problems with the code (or the product). These problems qualify for the definition of *side effects* since they are accidents that happen as a result of one of the main sources of benefits—the frequent builds. The process of changing and compiling the code on a daily basis is more likely to cause issues and breakdowns with the existing features. To address this matter, we need to further increase the amount of testing being done by running a series of tests against each build. While this course of action is likely to provide a remedy here, it is easier said than done since our testing resources are already strained. Another approach that can help relieve this issue is the incorporation of automated testing. This requires the need to select a tool and then automate the test cases. This is one place where the BA can help—directly and indirectly—the testers achieve economies of scale and ensure the processes are streamlined and efficient.

Less Documentation

One of the main benefits of agile's focus on efficiencies is reducing documentation. While the benefits of reduced documentation are clear and widespread, it may lead to an increase in the chances of error in capturing the intent or the acceptance criteria. It will put a strain on the entire team, but in the context of testing, it will lead to more pressure on the QA team since the problems may be discovered and realized only upon completing the testing. Direct involvement by the testers in articulating acceptance criteria (which also means more and earlier involvement of the testers), along with a BA who can ensure there are no gaps between the customer, the developers, and the testers, can help reduce this side effect to a minimum.

Testing Cycles Are Compressed

The iterations are compressed cycles of work where all team members work at capacity. There is a strong need for focused prioritization from both developers and testers; and with the time constraints, the testers may end up squeezing out some of the tests, especially when time is about to run out later in the iteration. If it is not done properly, the squeezed-out tests may be important ones that result in problems that are harder to fix later—close to the time of delivery. Naturally, this will further increase the amount of pressure the team is under, creating a snowball effect where more poor decisions will be made. A BA can help ensure that team members follow processes. They can serve as another set of eyes to ensure that no important tests are being squeezed out and they can be there when things go sideways to add capacity to the decision-making process at the late stages of the iteration.

Requirements Churn

Since requirements and code may change throughout the iteration, it may be challenging to ensure full test coverage. Testers must ensure that they do not miss new code that needs to be tested and that they test what needs to be tested thoroughly. Sufficient test plan documentation must be in place that will trace the test cases effectively to user stories to ensure proper test coverage (did we say that the BA can help with this?). Tools must be in place to ensure developers tag new source code so the testers do not overlook them.

The changing requirements (along with the continuous integration) may also cause the testers to miss certain important tests for some of the requirements, leading to test coverage issues. Linking tests to user stories for better insight into test coverage and analyzing metrics to identify traceability and missing test coverage are the best ways to address test coverage challenges. The BA can help ensure proper traceability by leading this effort, as well as helping team members follow the process to maintain a sufficient level of traceability.

Waterfallinizing

Agile methodologies and practitioners alike overwhelmingly reject the linear processes brought in by waterfall. However, many agile teams perform the development and testing work within iterations in a sequential fashion, making an iteration resemble mini-waterfalls, or *waterfallinize* the agile project. The sequential work between the developer and the testers increases the pressure on the entire team, compromises the benefits the agile project can produce, and strains the testing team. It is critical to *be agile* in the sense that the agile processes must take place within each iteration so the iterations do not become mini-waterfall cycles. This has to be done through seamlessly integrating the testers with the developers' work from the very start of the iteration.

Early Detection of Defects

Defects get progressively more expensive to fix the later in the development cycle they are discovered. The Cost of Change Curve[6] that is shown in Figure 8.3 illustrates several common agile strategies mapped to Barry Boehm's Cost of Change Curve.[7] In the early 1980s, Boehm discovered that the average cost to address a defect rises exponentially the longer it takes to find it. If we find defects minutes after they are injected, the fix will cost next to nothing; but if we find the defect months later, it may end up costing in a multiple of hundreds or thousands of dollars more. The reason for the exponentially higher cost is the

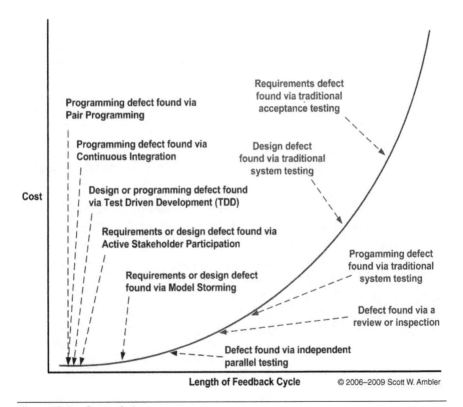

Figure 8.3   Cost of change curve

need to fix both the original problem, as well as all work that was based on the work that initially had the defect. To solve the issues that result from the longer fix cycles, the team needs to work collaboratively and perform frequent code reviews to identify and intercept issues early.

Sufficient Application Programming Interface (API) Testing

API testing involves testing APIs as part of integration testing to ensure that the interfaces' functionality, performance, and security are working properly. It is easy to overlook API testing due to the complexity of doing it—especially since many testers lack the coding skills that are required to perform it.

The Growing Role of the Testers

Testers are often viewed as gatekeepers of quality, getting involved to ensure that the product works as required before it is delivered to the customer. This is part of the linear way of doing things, and it is a natural progression of the way teams do their work: design-build-test. Although it is a logical sequence, it does not always work, and the late involvement by the testers is often not sufficient. Agile methodologies view the testers as full partners in ensuring quality, and as such, the testers need to be involved early, seamlessly, effectively, and proactively in the process—from defining the requirements and ensuring clear acceptance criteria is identified, to building and executing test cases. The agile view of the testers is new and serves as an adjustment to many. It breaks down organizational boundaries and by having the testers crossing over those boundaries, it may create *territorial* issues and friction between the newly *enhanced* role of the testers and other roles within the team.

A Waterfall Contribution to Fixing Agile Testing Issues

Not every practice that is done in waterfall environments is bad. Agile projects should *borrow* from waterfall the concept of an overall, all-encompassing test plan that includes traceability for all test types. Proper test plans should be developed early on (during iteration 0) and not become a last-minute rush just before each iteration. Similar to the project backlog, the test plan starts at a high level and will subsequently undergo thorough review and update cycles during each subsequent iteration.

Agile requires significant adjustments from the testing team to ensure sufficient test coverage, efficient build tests, and non-functional testing are performed effectively throughout each project in each iteration. Proper and efficient use of documentation, tools, automation, and reporting is key for project success—in agile, as well as in waterfall projects.

Agile Testing Best Practices

After reviewing common challenges with agile testing and ways to address them (some reactive, others proactive), this section provides a series of agile testing best practices. Along with the best practices, this section discusses the introduction of business analysis skills and the potential contribution of a BA to the agile testing effort. With so many ideas out there that suggest what the role of the agile tester should be, the confusion around the role of testers and how they fit with the team reinforces the need for business analysis skills to support the role of agile testers. It is safe to say that the role of the testers is the most disrupted role on agile projects. It is partly due to the growing and earlier involvement of testers in the building of the product, partly due to the increase in the work volume of the testers (because of continuous integration and regression testing), and partly due to the confusion around how the testers collaborate with the rest of the team.

Have testers become part of the agile team, with no distinction between the development team and the testing team. The Agile Manifesto calls for a focus on individuals and interactions—and this is accomplished by having one cohesive and cross-functional team without labeling the different roles. This means that testers are members of the team and they engage with others within the team regardless of titles.

As far as we can see in the history of product and software development, testing activities naturally happened at the end of the development cycle to ensure that requirements were met and that the product worked as intended. This is simply a progression of the actions we need to take in order to produce a working product since, obviously, testing a product requires it to first *be built*. However, the role of testing is typically associated with quality control and often leads to the testing process taking place too late after the development and build activities are done. This is reactive by nature and any defects that are discovered in this process need to be fixed by recalling the developers to the table and risking them producing side effects (breaking other things) while fixing the problems discovered by the testers.

Testers' Roles

As part of agile multifunctional teams, testers may wear many hats and perform many activities that are a direct part of the testing process, related to the process, or simply supporting it. All these activities are part of helping the team maximize the value produced for the customer and may involve activities that are beyond the traditional job description for a tester—demonstrating skills related to requirements, coding, and coordination. With such a variety of activities, it becomes quite obvious that of all roles in the agile project team, testers need to act as part BAs or receive the most support from the BA.

The following paragraphs list several roles and activities that come to mind when we think about the agile tester.

Testing

This is self-explanatory: testers need to write and execute test cases to ensure product requirements are met and that the product works properly. In the process, they need to find defects and bring them to the developers' attention. Testers must be proactive in their approach because they essentially provide a service for the team. Testers are a center of knowledge; they engage, listen, advise, and when needed—challenge.

Test Architecture

With design taking place virtually all the time (and in the form of fast tracking: design activities that take place in the current iteration are performed for the next iteration), testers play the *devil's advocate* to ensure that the design covers all the possible actions that a user can take and all possible related scenarios for the product. When something does not add up, the testers should bring the issue to the attention of those involved in the design process in order to improve the product's design. Any clarifications that are provided to the testers in the design process serve as an input to be used when testing takes place. A BA can help capture the findings from these meetings and consolidate information that can point out ideas for design improvements.

Explorer

Exploratory testing becomes very helpful in the effort to reduce risk. Exploratory tests are unique and nonrepeatable in the sense that testers perform them on a one-time basis as part of looking into a specific risk. The need to investigate specific risks usually comes up as a result of discussions related to stories or communication and collaboration activities within the team and with various stakeholders. Both sources can be facilitated, engaged, supported, or led by a BA. Sometimes a simple conversation may bring up some red lights due to concerns, unclear issues, or inconsistencies.

The exploratory testing is a proactive approach to uncover assumptions and to address the chance for problems before the actual issues surface. Needless to say, exploratory testing, which by nature is random and unstructured, must support a test plan and be part of a test strategy. As the exploratory testing proves the suspicion to be an issue, the testers need to resort to the more systematic approach as part of the test plan.

There is one more consideration to take into account in relation to the unstructured nature of exploratory testing: there is a strong need for coordination, resource allocation, and information management. These areas can be coordinated and supported by the BA because testers should not just retreat to perform exploratory testing without a valid reason or without proper coordination.

Automation

Not all testers need to do automation. The team needs to decide which tests need to be automated, and based on the tools selected and on the testers' and developers' relevant skills in doing automation, it will be determined who builds the automated checks. The BA can get involved in virtually every step along the way—help the search for an automation tool, determine which tests to automate, and coordinate the building of the automation.

Reporting

Test automation tools provide extensive reporting options, but since not all test cases are automated and some reporting needs go beyond what the tool can report on, the agile testers need to share related information beyond just reporting on the passing or failing of tests and uncovering issues and defects. In addition, the information sharing process has to be complete, consistent, timely, and as required according to the test plans. It is also important that testers provide reports on what they did not get to test, recurring issues and defects, blocks, and any other issues that stand in their way of performing the work. The reporting is only part of the extensive amount of communication that agile testers have with the rest of the team and, when required, testers should seek the support of the BA in streamlining the reporting and communication.

One of the indications of the effectiveness of agile teams is the lack of surprises. When things happen, the team communicates and collaborates on a timely basis to ensure that issues are addressed and things are not a surprise.

Testers have an important role in ensuring this. The use of defect tracking and reporting tools is only part of the tasks that testers need to perform since before logging defects, testers need to engage team members as part of an effort to clarify assumptions, remove uncertainty, and investigate. Many of these tasks are heavy on the business analysis side—and here too, testers may need to engage a BA to help enhance their communication with the rest of the team.

DoD

Testers need to ensure that the team comes up with a clear DoD. While it is not the testers' responsibility to define the DoD, it is often the testers' responsibility to monitor the work being performed by the team and ensure that each completed user story meets the benchmark DoD. This should be done starting with each new user story.

Collaboration on Requirements and User Stories

Testers must be involved in engaging and collaborating with the customer and other stakeholders for requirements so they can have a clear understanding of the acceptance criteria as part of the effort to create test case scenarios from these requirements. Proper understanding of the requirements also helps testers identify and capture complex and negative test scenarios.

Estimating Stories

Testers can help the team produce realistic estimates (e.g., by considering alternative scenarios) as part of the story estimating process. The value that testers bring to the estimating process goes beyond the estimates that they provide for their portion of the work. They also provide insights and clarifications to the team when certain possibilities or potential scenarios are overlooked.

Testing Goals and Vision

The testers, along with the Scrum Master and the BA, share the task of ensuring that the testing vision and goals remain visible to the team. Testing is often viewed as a boring, mundane, and frustrating chore; and it is important to ensure that team members view the testing effort as a legitimate, important, and dynamic part of the team effort.

Participate in Demos

Due to the short iteration cycles in agile projects, developers often have less time to explore the complete functionality of their user stories on their own. This leads to the need of enhanced collaboration between developers and testers since the developers need testers to provide clarifications and share their familiarity with the complete functionality of each story, along with the acceptance criteria. Many agile teams get the testers to not only participate in the demo at the end of the iteration, but to lead the product demo at the iteration review. With the comprehensive knowledge of the stories and their acceptance criteria, the testers can provide answers to many of the questions and inquiries that the business stakeholders may have. This can help the developers focus on the more technical questions and issues that come up during the demo.

Keys for Success

The agile testing best practices that are being discussed here are the building blocks of agile testing success. With the team in agile projects being a cross-functional, cohesive one that includes the testers, testing must take place in real time and be a continuous process that is proactive and encompassing, rather than being viewed as a separate phase. Testing is also a process that has virtually everyone on the team involved. It is not done exclusively by the testing team. In addition, the feedback that the testers provide is quick and continuous (within the same iteration in an effort to minimize technical debt), which helps to reduce the cost and response time to fixing problems. Efficiencies do not skip over the testing effort since agile testers minimize documentation, utilize checklists, and incorporate efficiencies through scaling the testing processes via automation.

Back to the BA and the Testers

Earlier in this chapter we looked into some of the touchpoints between the roles of the testers and the BA in agile environments. Now, after reviewing the wide range of responsibilities and tasks that the testers perform, we can see even more overlap, collaboration, and almost a role-convergence between the BA and the testers. With the testers responsible for defining and refining the acceptance criteria for the stories, testing the iteration's functionality, and doing regression testing, we found multiple additional touchpoint opportunities between the BA and the testers. While the BA actively supports the testers' efforts, that same support helps the BA facilitate the backlog maintenance and prioritization process, oversee story analysis, and consolidate feedback and communication with the product owner.

But the BA Is Not a Tester

Despite the close collaboration between the BA and the testers in an agile environment, it does not mean that there is full convergence of these two roles. Testers need to have business analysis skills in order to perform the wide range of tasks that are necessary in support of the team and to maintain the quality and integrity of the solution. As the testing effort requires an increasing amount of help (especially from the BA), it is common to see BAs get more progressively involved in supporting the testing effort; and in many environments the extent of their support turns into having the them participate in the actual testing. While the testing help that the BA provides is valuable, turning the BA into a tester is not the best way to utilize the skills and experience that he or she brings to the table.

This is also true when looking at it in the opposite direction: testers should not be the BAs of the agile team because there is a need to have devoted testers for the agile project and, therefore, separating both the business analysis and the testing from the development work is the right thing to do. With that said, it is highly beneficial for the testers to have business analysis skills and for the BA to be able to help with the testing effort, but experience shows that separating these roles is the way to go.

AGILE TESTING STRATEGY

The goal of having a test strategy in agile projects is to provide a list of best practices along with a structured foundation for the team to follow. As with other things—such as the need for governance in agile projects—agile does not mean that things are unstructured. The BA can get involved in putting together a test strategy in multiple ways—from leading the effort to providing insights, access to information, or advice. An agile test strategy covers the following areas:

- *A mission statement*: that mentions the intent and importance of QA and the testing process
- *Test levels*: types of tests—such as functional, stories, prototypes, simulations, unit testing, component, performance, security, non-functional, exploratory, usability, API, system, regression, and user acceptance testing; this includes a breakdown of roles and responsibilities for each test type and logistical information
- *Guidelines for automation*: such as automation approach, tools, coverage, types of test cases to be automated, and the number of automated cases
- *Guiding principles for what makes a good user story*: format, Bill Wake's INVEST criteria for good stories; unclear requirements and misinterpretations of requirements in the backlog are a leading cause of issues, problems, defects, and product failure
- *DoD and definition of acceptance criteria*: ensuring that each story has clear acceptance criteria that is embedded with the story and written at the same time the story is created
- *Guidelines*: for story creation and workshops

RECAP

Of all the roles on an agile team, the timing and work volume of the tester can change the most. In this chapter we reviewed the agile approach to quality, the role of the testers and the testing activities, as well as challenges with and best practices in agile testing, including testing automation considerations.

Throughout the chapter, we continuously explored the touchpoints and connections between the roles of the BA and the testers, along with new ways where the BA supports the testing process. While it is common to see BAs getting heavily involved across the testing process—including, in some environments, actually doing the testing—the need for the BA to support the testing process is clear, but not by becoming a tester. When looking at the testers' role, we established that it is almost a necessity for the testers to have business analysis skills in order to enhance project success through proactiveness, collaboration, communication, and involvement in the requirements, acceptance criteria definition, and user story estimates. With that said, it is not recommended to have full role-convergence between the BA and the testers since this will not satisfy the intent behind the two sets of roles. The BA must continue to provide support to the testers, the testers must have a multitude of BA skills, but where there are BAs in agile projects, their roles should not combine with that of the tester.

Testing has a key role in delivering project success and value to the customer, and it is critical to maximize the contribution of the testers to the team's effort through the introduction of best practices and testing strategies, and how we define BA support for the testing effort.

ENDNOTES

1. In this chapter, we will refer to the testing team, QA, QA team, or testing analysts interchangeably.
2. Winston W. Royce. (1970). "Managing the Development of Large Software Systems" in: Technical Papers of Western Electronic Show and Convention (WesCon) August 25–28, 1970, Los Angeles, CA, U.S.
3. ibid.
4. Adapted from http://www.modernanalyst.com/Resources/Articles/tabid/115/ID/1967/A-Proposal-for-an-Agile-Development-Testing-V-Model.aspx.
5. Adopted from Agile Alliance, https://www.agilealliance.org/glossary/exploratory-testing/#q=~(filters~(postType~(~'page~'post~'aa_book~'aa_event_session~'aa_experience_report~'aa_glossary~'aa_research_paper~'aa_video)~tags~(~'exploratory*20testing))~searchTerm~'~sort~false~sortDirection~'asc~page~1).

6. Adapted from http://www.ambysoft.com/essays/agileTesting.html.
7. http://csse.usc.edu/new/barry-w-boehm.

9

A DAY IN THE LIFE OF AN AGILE BUSINESS ANALYST

While understanding agile theory is helpful, looking at how theory is put into practice can help business analysts (BAs) understand where and how to start implementing agile techniques. This chapter describes what a typical day looks like for an agile BA and how the BA can add value while working together with the rest of the agile project team. When embarking on the task of operating agile projects, many teams will have the necessary education, theory, and concepts in place, but will lack the practical application of these concepts and how they translate into day-to-day activities. Further, with the fast pace of events and activities in agile projects, team members often get lost in the variety of ceremonies and interactions that take place.

Similar to many other fields, the most effective way to approach the challenges that the agile daily routines pose is to create patterns that help articulate, define, and sequence things. This will help the team understand the overall construct of a day and to plan and set expectations accordingly. It will also help the team fully understand its own capacity since blocks of time are required as part of the agile processes. In fact, even agile practitioners have misconceptions about the need to perform many of the agile ceremonies and planning. Quite often, people refer to the activities that take place on the first and last days of the iteration as wasted time. They, in turn, attempt to minimize the amount of these *overhead* activities, and by that, compromise the quality of their project work. These *overhead* activities include the iteration review and demo, the retrospective, iteration planning, prioritization, and other elements that are the backbone of agile methodologies.

When teams and organizations try to artificially reduce the amount of time they spend on these *overhead* activities, they resort to more aggressive scheduling practices where they try to force more working time into the iteration. For example, practitioners try to achieve a total of five days of development in a week-long iteration. It is obviously not possible to achieve this because there are planning, prioritization, troubleshooting, testing demos, and other activities that need to take place. The aggressive scheduling practices inevitably lead to performance and quality issues. Alternatively, in waterfall environments, there are also many interruptions that add up to a lot more time—and often turn into distractions. There are actually less interruptions in agile projects than in waterfall environments, but it has to be articulated that following agile methodologies does not mean that the team has *carte blanche*. There are still activities and ceremonies that need to take place. Overall, the *overhead* interruptions, agile ceremonies, and planning activities (or whatever people want to call these things) add up to significantly less distraction and meddling than in other methodologies.

DAILY PATTERNS

When agile teams struggle with their daily routines or try to reduce them below the amount of time they require, it goes beyond just challenges in managing stakeholder expectations: it often translates to performance issues, reporting challenges, and ultimately—quality problems. Since there are events and ceremonies that need to take place for practical reasons as part of keeping the agile team going, when these activities are not performed to the extent needed, agile teams often find themselves struggling to meet their commitments. The arrangement and planning of the blocks of time that involve the *overhead* activities that we will focus on in this chapter fall under a combination of three items: a checklist, a schedule, and an algorithm.

- *The checklist*: The agile processes and best practices require the team to come up with a list of work products, processes, deliverables, and activities that need to take place throughout each iteration of the project. These activities are repetitive from one iteration to another.
- *The schedule*: The next step is to sequence the activities and create blocks of time. First, identify how much time the team should allocate for each item, and then consider activities and escalations that may need to take place and position them where they belong during the day. The agile team needs to keep in mind that while Scrum provides very specific time-box durations for each activity and ceremony, the team should identify blocks of time that are according to the team's needs and dynamic. The BA can facilitate the process of creating these checklists and

support the building of the *day schedules* to ensure that no activities are forgotten. Further, the BA can work with the project manager (PM) or Scrum Master to monitor the team's performance around the daily routines and propose improvements along the way.

- *The algorithm*: This will help the team make the schedule more modular so it is broken down into two parts. (1) Things that the team needs to do (e.g., iteration planning meeting, demo)—these things generally have a preset amount of time allocated to them, they take place in every iteration, and they are easier to identify and plan around; and (2) things that the team may need to do, but (with the help of the BA) team members identify the trigger conditions and the potential turnaround time in case these events occur (e.g., dealing with impediments that were discovered in the daily stand-up meeting). These things are conditions so they may or may not happen. Further, if they take place, they may fluctuate more in time based on the extent of the issue they are addressing. With that said, these events are still possible to plan for. In this case, the team (facilitated by the BA) identifies the trigger conditions to look for; and if they are present, then the associated activities or events will take place. The BA does not need to facilitate the actual events of dealing with the impediments (unless the BA has a specific role in this process), but the BA helps the team keep an eye on the presence of the trigger conditions. The retrospective at the end of the iteration will also track the frequency, severity, and impact of those triggered events for further identification of triggers and for potential improvements in handling these events.

Distinct Days

These mechanisms help the team design their days in the agile project. Further, agile projects can be broken down into three types of distinct days:

1. The last day of an iteration
2. The first day of an iteration
3. Any other day throughout the iteration

The rationale behind the breakdown is that there are three unique days for an agile team throughout each iteration. Most of the decisions and due-diligence activities that lead to a new iteration take place and peak on the last day of the previous iteration. While not every agile team follows the exact same practices, the scenario described in this chapter outlines an approach common to many experienced agile practitioners.

Agile teams agree to their *day plan* every day during their daily stand-up meeting, which is typically held first thing in the morning. The day plan is established from the answers to the second question asked during the daily stand-up meeting: "What are you going to work on today?" Since the day plan changes every 24 hours, there is no point in documenting the day plan—it will be obsolete by the time the next day rolls around. Day plans vary by industry, organization, team, and project type. Essentially, they need to be established by the team on a project-by-project basis to account for unique project and team circumstances. Considerations include factors such as the daily start times of each team member, what other commitments (multitasking) the team members are supporting, and the timing of competing meetings. The BA steps in where team members struggle to account for all considerations, as well as for coordination purposes.

Differing Levels of Formality

Because business stakeholders (including the product owner) need to be involved in certain iteration start-up and shut-down activities (e.g., end-of-iteration demos), certain activities need to be formally scheduled to get time on the business stakeholders' calendars. Also, meeting rooms may need to be booked for these activities. As a result, teams often establish a formal *template* for the first day of an iteration and the last day of an iteration—where these formal activities take place.

A typical middle day of an iteration, however, is normally quite flexible—other than the start time of the daily stand-up meeting, the team may have no other formal arrangements. The activities of middle days are normally agreed to during the answering of the second question of the daily stand-up meeting.

SAMPLE CASE SCENARIO

Tyler and his team are working for a technology start-up company that is funded by third-round venture capital financing. They expect that their IPO is about eight to twelve months away and have a nearly complete product with interested customers already lined up and eager to buy. Their product is a system that has computer-controlled drones patrolling parking lots with license-plate recognition software. The purpose of the system is to better manage the parking lot for the property owner by improving security, better managing the handicap parking spaces, and preventing unauthorized parking. Fully automated drones can capture license plates from parked cars and run them through the local department of motor vehicles database. Results can be used in multiple ways:

- Capturing car license plates and possibly driver or passenger photos in order to identify who has been on the property in an effort to discourage car theft and vandalism
- Determining who has an authorized handicap parking permit so that if a car is illegally parked in a handicap space, then the software can notify the parking authority and the towing service to ticket and remove the vehicle immediately
- Determining if any cars have been parked in excess of the maximum permitted time in order to prevent workers in nearby offices from parking illegally in a local shopping mall; if an organization provides the license plate numbers of those with parking permits, then the system can notify a tow truck to remove any unauthorized vehicles not in a visitor parking space

As a bonus for law enforcement, the system optionally notifies police if license plates show up for:

- Drivers who have not paid court-ordered child support payments
- Drivers whose licenses are under suspension
- Cars whose license plate registration has expired
- Sought-after cars due to amber alerts, car theft, arrest warrants, or other reasons

Tyler is the BA on the agile project to design and build the solution. There are six more one-month iterations to complete the solution and have it ready for deployment with the first customers. So far, a local shopping mall operator, the property manager of a local office tower, the owner of a medium-sized strip mall located across the street from a high school, and a large multiplex movie theater are signed up as customers who will participate in the initial pilot of the solution. Tyler has worked with each of these initial clients to capture their high-level requirements and has helped prioritize them with the other functional requirements that were considered for the solution. For this case scenario, let's start by looking at the last day template for the next iteration.

The Last Day Template

The template for a typical last day of an iteration looks quite different from the first day's activities—or any other day's activities. The main outputs of the last day's activities include the completed stories and features, the updated project backlog, and the documented lessons learned. These items also help shape and finalize what will take place in the new iteration on its first day. Here is a list of the major activities that take place on the last day of the iteration:

- Hold the daily stand-up meeting
- Clean up the deliverables and prepare them for the end-of-iteration demonstration
- Rehearse (do a *dry run*) of the demo
- Perform the demo, capturing stakeholder feedback
- Perform the retrospective (lessons learned) workshop and document the results

In this case scenario, the team has decided to hold the daily stand-up meeting at 9:15 in the morning, giving team members 45 minutes to get settled and organized after their 8:30 a.m. arrival. As a core team member, Tyler participates every day in this stand-up meeting, asking for help from the team for any of his current issues and identifying any new risks he sees that may impact the chances of the team meeting its commitments. After any resulting follow-up discussions, Tyler gets involved in planning the demonstration. He will likely be running the client through the demonstration since he is the one on the team who best understands the client's nuanced needs and can talk to the client using the client's own business terms, mapping the demonstrated items back to the project's business case. In planning the demo, Tyler does not demonstrate every new/revised feature; rather, he focuses on three things:

1. *Feedback*: At times, there are areas where Tyler and the team need additional feedback to clarify a requirement or to confirm whether or not they have met the client's expectations for the requirement.
2. *Requests*: These would be for items that the product owner or other business stakeholders have asked to be (re)demonstrated in order to help the requestor understand the solution better.
3. *Boasts*: This involves items that the team wants to demonstrate to the product owner or business stakeholders in order to *show off* what they can do—expanding the client's understanding of what is possible ("Wow, team, that is incredible! I had no idea you could do that. It made me think of a way to improve the value of what we are delivering . . .") and to boost team morale by receiving positive feedback from the clients ("Way to go! That's amazing. I didn't know you guys could do that.").

Next, the demo is held with key stakeholders. During the demo, if the business stakeholders have any feedback for the team, it is likely in the form of new features, changed features, defects, or other items that may affect the vision, business requirements, design, or other elements. Capturing this feedback leads to adding the information as new items in the product backlog, defect backlog, or as process improvement ideas to use when planning the next iteration. After the end-of-iteration demo is completed, Tyler participates with the rest of the core team in the retrospective (lessons learned) discussion, identifying both successful and unsuccessful behaviors and techniques, creative ideas for further performance enhancement in the next iteration, and solution areas that may need refactoring (rework).

After the retrospective ends, the iteration formally closes. Tyler is likely still going to have some time left in the day. He can be productive with this time not only by catching up on e-mail and other non-project administrative chores, but he can also perform some project-related work, such as helping with backlog maintenance (a.k.a. backlog grooming), where existing backlog items are reprioritized and new backlog items are estimated and inserted into the backlog in order of priority.

Timing

Beyond understanding what needs to be done during the last day of the iteration, there is a need to break the day down properly in order to ensure that everything that needs to be done fits into the day. The BA, along with the Scrum Master, should provide the team with guidance and support with putting these *schedules* together. The goal is to work with the team and identify time blocks for all things that need to take place during the day. These blocks of time are not *laws*, but rather guidelines that may, or rather should, change throughout the project—based on the actual needs of the project and the ability of the team to support it. Tyler believes it is right up his alley to facilitate the logistics of these time blocks and to stay on top of the need to incorporate all feedback, trends, and needs—to fully reflect the team's and the project's needs. Similar to all other administrative and process-related issues to be brought up in the retrospective, any need for changes in the day's timelines should also be reviewed and considered in that meeting, which is the last structured part of the iteration's last day. The time breakdown, as reflected in Figure 9.1, goes beyond helping team members budget their time accordingly; it also helps manage stakeholder expectations and provides them with stability and consistency so they can provide input and make themselves available.

The BA

Whether it is a BA or team members who act as BAs, Tyler is vital when it comes to tying up loose ends, ensuring that team members are prepared for

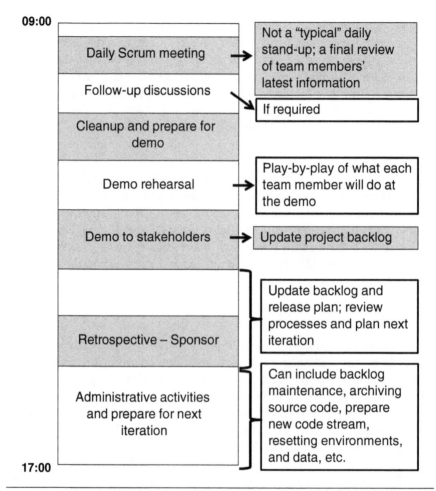

Figure 9.1   A typical last day template

their role in each activity, and confirming that all activities are performed properly—including:

- *Stand-up meeting and follow-up discussion*: Although Tyler does not have a specific role in the daily stand-up meeting, he gets into action in the follow-up discussion to ensure any items that require follow-up or action are addressed and that there is sufficient access to the relevant stakeholders and subject matter experts (SMEs).
- *Cleanup*: The BA is also useful in connecting between team members and other stakeholders in the final cleanup before the demo.

- *Demo rehearsal*: This session has to be facilitated just as if it is the real demo and the BA is often the one who facilitates it, depending on the breakdown of work between the BA (Tyler), and the PM/Scrum Master (Skylar). The BA also prepares during the rehearsal—taking the right notes and capturing the demo's feedback so that all information is reflected in the project backlog.
- *Demo*: Tyler's role in this session, as well as all other team members, will be to follow the path that was developed during the rehearsal.
- *Retrospective*: There will be two retrospectives—the first one for the team only and the second one with the sponsor. This breakdown allows the team to have a candid and open discussion and to bring up *raw* information from the iteration and their work. Tyler, the BA, should capture the key action items and track them throughout the next iteration to ensure that they are incorporated properly. At the next iteration's retrospective, the BA will present the findings on the application of these lessons. Though before moving ahead with any recommendations and calls for action, the second part of the retrospective will take place—this time with the sponsor. It allows for a more focused discussion and it respects the sponsor's time since the team arrives at this session more prepared and with a clearer set of findings and recommendations. The sponsor will review these items and provide the approval for the proposed actions. This flow makes sense for many agile project teams, but some agile teams conduct the sessions in reverse order—having the retrospective with the sponsor first and then the team-only session. There is no one clear and correct way to do it so it is up to the individual team to decide that. Either way, the format of splitting the retrospective into two sessions works effectively and it provides a much better use of both the team's and especially the sponsor's time. It also allows the team to produce a more focused and clear list of action items for improvements.

The First Day Template

The template for the first day of an iteration requires the team to produce an updated release plan, iteration plan, and task estimates—performing a number of activities:

- Backlog maintenance, if not completed at the end of the last iteration
- Confirming scope of the new iteration
- Updating the release plan, if not completed at the end of the last iteration
- Explaining the vision for the new iteration

- Explaining the requirements and the solution design that was prepared for the last iteration to those building and testing the components in this iteration, if following the agile fast-tracking approach
- Preparing estimates and planning the new iteration

Although there are many things listed here, the activities on the first day of an iteration do not add up to a full day. When the team completes all of these activities, team members will move to the next item on the list:

- Start working on the scope of the new iteration

Fast-Tracking

As part of the fast-tracking approach, the design work for an iteration takes place during the previous iteration, where each iteration acts in part as Iteration 0 for the next iteration. The rationale behind this is to allow the work in the next iteration to begin as early as possible in the new iteration—and it is enabled by having the design work completed (at least for the most part) one iteration in advance. Although the formal feedback received at the end of the iteration and the demo are the main contributors to the process of the backlog maintenance, the team will know, primarily, what items need to be included in the next iteration based on the events of the current iteration. This knowledge allows the design work to proceed and saves time in the next iteration.

The performance of the team, the rate at which the team makes progress, the team's velocity, and the type and severity of defects produced will all serve as input into the process of determining what is likely to be included in the next iteration. In addition, any new and emerging needs that stakeholders may introduce throughout the iteration do not often wait for the end of the iteration and are usually introduced in real time during the iteration. The fast-tracking technique is not perfect since it introduces some calculated risks, but it provides the team with the ability to charge forward responsibly without taking any unnecessary shortcuts. The fast-tracking technique requires strong team cohesiveness, coordination, discipline, and trust within the team and between the business stakeholders and the team. Tyler, like other BAs, is in a key position to ensure that all considerations are taken into account as part of the fast-tracking process, including the introduction, documentation, and validation of assumptions. Similar to the *traditional* fast-tracking schedule compression technique, there are risks associated with it; the main ones are introduced when the team is careless, when there is no assumption management, and when the trust between the business stakeholders and the team is compromised.

Validation

- Preparing the iteration backlog and iteration plan for the upcoming iteration
- Committing to the scope (goals) of the iteration

Tyler and the team would get involved in backlog maintenance and with updating the release plan *only* if these activities were not completed on the last day of the just-finished iteration; otherwise, they do not need to be duplicated at this point.

If an agile fast-tracking approach is being used, where the requirements analysis and design work are performed one iteration ahead of the related build and test work, then Tyler would get involved with explaining the detailed requirements and design for the deliverables or solution features that are expected to be completed in the new iteration. Tyler may spend the first couple of days explaining or answering questions about this iteration's requirements and design to the builders and testers. Once they are happily working away on building and writing test cases for the new solution elements, Tyler and the rest of the requirements/design team will move ahead on gathering detailed requirements and performing detailed design activities on the next iteration's scope items. The last few days of the current iteration will see Tyler and his peers supporting the testers by clarifying business requirements and confirming that the new solution components are working as designed.

On the first day of a new iteration, Tyler will also get involved in helping the team figure out how they will get their work done in the iteration. This means that Tyler will provide the Scrum Master or PM with the work breakdown structure—or a detailed task list—for capturing the detailed requirements (plus any related models) for the current or future (if using agile fast-tracking) iterations. In addition to the tasks, Tyler will also have to provide detailed estimates—in hours or days—for the tasks in the iteration backlog.

Timing

Here, too, in a similar fashion to what happens on the last day of the iteration, the blocks of time are not *laws*, but instead guidelines that may, or rather should, change throughout the project—based on the actual needs of the project and the ability of the team to support it. With the first day focusing on finalizing the work for the new iteration, Tyler will continue to be useful in addressing gaps and providing clarifications, as needed. The time breakdown, as reflected in Figure 9.2, shows the activities' blocks of time throughout the first day:

- *Backlog maintenance and updates to the release plan*: If there is a *spillover* from the last day of the previous iteration that requires more work toward finalizing these two items for the new iteration, team members will engage in these activities. Otherwise, the focus shifts to the main goals of the iteration's first day—to get the new iteration going.

- *Confirming vision and scope for the new iteration*: This activity is led by the product owner to ensure that everyone is on the same page for the new iteration. With the feedback from the demo incorporated into the backlog, proposed changes approved, and the backlog maintenance activities complete, one would argue that there is no real need to clarify the vision and the scope for the new iteration, but with all the moving parts and with potential changes that have come up about the new iteration,

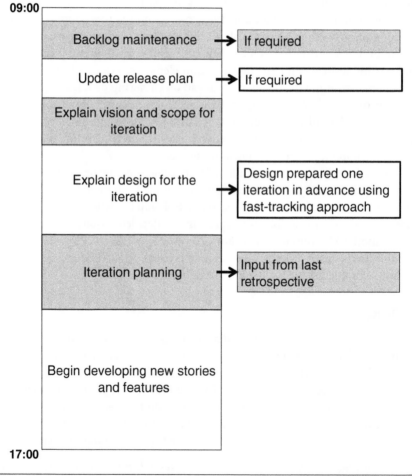

Figure 9.2 A typical first day template

it is important to get the iteration's theme, vision, and scope clear and to answer any questions that team members may have.

- *Explaining the design for the new iteration*: The designers who completed the design work in the previous iteration need to articulate and explain the design to the team.
- *Preparing estimates and planning for a new iteration*: With all the information now clear, team members will prepare (or finalize) their planning for the iteration, including the breakdown of tasks and estimates.
- *Starting work on iteration scope*: After all other activities are finished, the team will start working on the stories and scope for the iteration.

The BA

- *Backlog maintenance and updates to the release plan*: The BA supports any needs that may come up and ensures that all input is incorporated into the backlog to reflect the latest set of needs.
- *Confirming vision and scope for the new iteration*: This activity is led by the product owner, but the BA provides clarifications, asks questions, and captures the answers.
- *Explaining the design for the new iteration*: The BA took part in the design activities and provided support to the design team. The BA is very useful in ensuring that the communication between the design team and the rest of the team members is clear and that the information is captured properly.
- *Preparing estimates and planning for the new iteration*: Each team member performs planning and estimating for their stories, but there is a need that the BA can fulfill for clarifications, cross-story coordination, and other support to ensure that tasks are identified and that estimates are realistic.
- *Starting work on the iteration scope*: The rest of the iteration's first day looks like any other middle day of the iteration. It is expected that not all team members will get to this stage exactly at the same time since the estimating and coordination work may vary from one team member to another. The BA also helps provide final coordination on story dependency and any missing information, while also helping with the tasks and estimates.

The Middle Day Template

The things that happen on the rest of the days in the iteration will change depending upon where the team is in the iteration. For example, in the early days of the iteration, most of the work focuses on building new solution components, while in the later days of the iteration, most of the work focuses on finding, correcting, and verifying/retesting defects. There are a few activities that are common to middle days, though the emphasis may change based on where the team is in the iteration.

A typical middle day starts with the daily stand-up meeting, attended by all core team members (including Tyler) plus as many supplemental team members who wish to participate. Tyler's team holds the daily meeting at 9:15 a.m.—but it is not mandatory to hold the stand-up meeting first thing in the morning. Having this meeting early in the morning allows for a logical flow of events since the team members recap the previous day's work, answer questions about where they are with their work, and talk about the upcoming days.

Daily Stand-Up Meeting—In the Middle of the Day?

However, many types of circumstances may restrict certain team members' ability to make it on time for the meeting. In such situations, some teams resort to having the meeting in the middle of the day just before lunch. This kind of format slightly changes the team's timelines as now the *start* of the reporting day takes place in the middle of the day. In these situations, the team will arrive in the morning and members will continue to perform tasks that they started to work on during the previous day. The daily stand-up meeting will mark the end of the reporting day and the team will then start the activities of the *new* day in the afternoon. Some practitioners prefer this kind of cycle since it does not force team members to arrive early in the office; and with the meeting starting at 11:45 a.m., there is almost a guarantee it will end by noon.

After the Stand-Up Meeting

Following on the heels of this short meeting, key team members (again, including Tyler) may wish to stay behind and discuss opportunities for collaboration, mentoring, knowledge transfer, and coordinating their day-to-day work. After the daily meeting is finished, Tyler will get to his typical daily work running requirements-gathering workshops, documenting detailed requirements, and verifying requirements—often using models—with key stakeholders. In addition to the support that the BA provides to the design work for the next iteration that is taking place during the current iteration, there may be additional areas where the BA can provide support for the team (and for the Scrum Master).

Despite the changing nature of work from one iteration to another and from one day to the next, there are typical outputs that are expected, including corrected defects, completed scope items, and updated task estimates. It is still easy to identify clear patterns in the structure of the middle day's activities:

- *Fix blocking defects*: After the stand-up meeting and the follow-up discussions, there may be a need to fix blocking (high severity) defects. Only those who are involved with these defects will be part of this effort, including the team members whose stories face these defects and other team members (and SMEs) who are part of the solution. When facing blocking defects, there may be a need to reassign some of the stories to other team members based on their skills, experience, and relation to the problem. The BA, in support of the Scrum Master, may be required to assist in the effort of looking into the blocking issues, helping team members get assigned appropriately, and ensuring that the reporting reflects the project's progress.

The Rest of the Day's Timing

- *Continue working on scope*: Team members who are not involved in the defect repair will continue working on their scope of work and stories and on producing features. Once any defect repairing activities are complete, the other team members will also proceed to continue working on their respective stories.
- *Testing (when possible)*: Depending on the way the team performs the work, there may need to be unit testing or even daily build testing. Generally, toward the end of the iteration, the main focus of work will shift toward testing to ensure that the features, functionalities, and stories that are built perform to the extent that they should. Chapter 8 of this book covered the area of testing and the BA's role in supporting the testing effort.
- At the end of the day, team members will produce an update for the Scrum Master, so that Skyler will have a clear idea of the daily progress and the overall status of work at the end of the day and leading into the next one. While there is no clear role for the BA as the team members work on their scope items or provide updates for the Scrum Master, it is common to see the BA supporting many of these activities that lead to wrapping up the day—providing consistent updates and coordinating some of the testing effort. Otherwise, when there is no need for the BA to support the routine activities during the day, the BA will facilitate and support the requirements and design work for the next iteration.
- Each team member finishes his or her day by updating their task estimates based on their work progress and ensuring that newly found defects (if there are any) are documented, reported, and assessed as an input for the next day's stand-up meeting. Here, too, the BA provides context, access to SMEs, and an objective assessment of the new defects. In addition, the BA often supports the process of updating the task estimates that are shown in Figure 9.3.

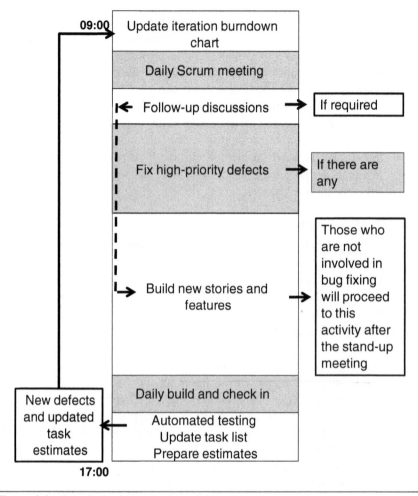

Figure 9.3   A typical middle day template

RECAP

Business analysis activities take place regularly throughout the project (and throughout the iteration) when requirements are added, reprioritized, and deleted. Requirements need to be elicited from key stakeholders, analyzed and documented, and then explained to those who will be designing, building, and verifying (testing) the solution. While there may be no formal BA performing this work, the role of the BA needs to be performed by one (or more) of the team members.

An agile iteration can be broken into three types of distinct days: the last day of an ending iteration, the first day of the new iteration, and any middle day of the iteration. No matter what the iteration's duration, the activities listed here need to take place. While some of these events, ceremonies, and activities are essentially mandatory as part of running agile projects, other events are optional based on the team's needs and progress. The extent of the events and how long they are will be determined based on how the team conducts itself and the needs of the project. While Scrum provides specific guidelines as to the duration of each event, the team—with the help of the Scrum Master and the BA—needs to determine how long each activity should last. Similarly, the role of the BA changes from one environment to another and from one situation to the next because the BA is there to support the team (and the Scrum Master), streamline the flow of information, provide clarifications, help escalate issues and identify courses of action, and assist in finding the right type of available expertise to address the issues. The BA also supports the design and requirements work that takes place during an iteration as part of the fast-tracking process for the next iteration, and the BA facilitates the process of backlog maintenance to ensure that the latest needs are reflected and addressed in order to maximize value for the client. Finally, the time blocks mentioned in this chapter are only guidelines—and although they try to be proportionate against each other, they do not call for a specific duration of each of the activities. Teams must keep in mind to contain the duration of the meetings, especially those meetings that involve external stakeholders.

10

MOVING FORWARD

It is safe to say that for the foreseeable future, agile is here to stay. Until a few years ago, some people thought agile was a fad; but this is not the case and agile is not going anywhere. However, as an increasing number of organizations begin (or at least, attempt) to apply some agile concepts, there have been and will be changes in the way we view agile. Many experts and practitioners view agile as almost the *ultimate* answer to many of the challenges we face in product development, innovation, and software development. They also view agile to be very suitable for the characteristics and thinking processes of millennials (we will discuss generational challenges and considerations later in this chapter), which implies that agile will continue to grow and serve our changing needs— almost perpetually.

However, people tend to fall victim to group think and fit into paradigms; similar to many analysts, who, with the lack of clear evidence, tend to predict that current trends will continue at the same trajectory as they currently are. With that said, every trend gets hit at some point by opposing and disruptive forces that interfere with its trajectory, introduce challenges, and lead to new needs and trends. The problem is that it is hard to see how the tide will change—just as it is hard to see during a stock market rally how it can possibly end. While agile is a pragmatic, flexible, and practical approach that addresses our current needs (as well as our past challenges that were part of the waterfall approach), this chapter reviews some of the things that happen within and around agile in a quest to identify if and when a disruptive element will be introduced into the mix.

Up until the time that agile was introduced, the mainstream way to build software (and other products) was waterfall. Since then, agile has increasingly become mainstream to the extent that some people find it almost embarrassing to admit that they still manage projects the waterfall way. But with the application of meaningful milestones and with the phased approach for introducing products, it becomes hard to find many projects that are managed in a *full* or complete waterfall way. Now that more and more organizations are attempting to go agile, we need to keep in mind—as we reiterate throughout the book—that not all projects (or customers, or environments) are suitable for agile projects. When there is little to no change in needs and requirements throughout the project, where the scope is clear from the start, and when there is no need for early delivery of value in a phased approach, waterfall may be sufficient, simple, and efficient to follow. In these cases, it would be a waste to employ all of the mechanisms and efficiencies that agile offers since there is likely no good use for them.

Yet, as the majority of projects follow agile concepts, there will be a growing number of challenges to address and many of these are related to, or can be at least partially addressed by, the role of the business analyst (BA) in agile environments. Further, since agile methodologies do not call for a person with a BA title within the team, as agile becomes increasingly mainstream it introduces some existential questions about the role of the BA and even the need of the BA in agile environments.

As established in this book, we can agree that there is a clear need for business analysis skills in agile projects. These skills can be performed by team members (implying that technical roles partially act as BAs) or by a dedicated BA. The questions around whether there is a need for a dedicated BA depend on the organizational needs and the team's ability to deliver those needs. A wide range of considerations need to be taken into account—including the following:

- *Team maturity*: This is measured by the ability of the team to work together effectively and perform the work along with the agile ceremonies and artifacts.

- *Team cohesiveness*: Team members need to effectively elicit and manage requirements (our apologies to all agile practitioners for using the word *requirements* in this book since in agile we have user stories) and build the right product to address the client's needs. Some teams do not have what it takes to be self sufficient and to handle customer needs without guidance from a BA and/or a project manager (PM).

- *Team members' skills*: When team members need to wear two hats (one for their respective technical roles and the other for business analysis), there is a need to check whether individual team members have the capacity to do so. Beyond the need for team members to share the application of business analysis skills, there is often a need for team members to be well-rounded with their technical skills (e.g., the expectations in some environments for testers to know how to write code) and generally well-rounded in understanding the big picture—with the ability to work on requirements, work directly with the client, and communicate effectively. When team members lack any one or more of these skills, there is a need for someone to step in and cover the potential gaps and the issues that may ensue—and this person would be a BA.

- *Organizational considerations*: An important consideration would be the availability of the product owner and his or her ability to provide ongoing, timely, and meaningful feedback and decision making regarding the work's progress and changing needs.
- *Other factors*: These could include the nature of the involvement of additional stakeholders, the extent of the changing needs, the extent of processes, compliance, and red-tape in the organization, the impact from other projects, resource allocation, governance and other organizational policies, and environmental factors.

THE CHALLENGES AND THE BUSINESS ANALYSIS SKILLS

There will always be a need for business analysis *skills* in projects. Whether these skills come with an actual BA or are shared among team members may differ from one environment to another; it depends on multiple variables. With the rapid growth in attempts for agile adaptations, it appears that the need for an actual BA is going to trend upward. Agile's growth across industries and into organizations that were not previously considered as the most obvious candidates to run an agile project (i.e., large financial institutions, government agencies, and other large, unadventurous, and slower-moving entities) introduces a slew of challenges, issues, and problems.

In many waterfall environments, the BAs are there to cover for the deficiencies of team members, stakeholders, and processes; while in agile projects, where presumably many of the waterfall-associated deficiencies are fixed, the BA (or a team member with business analysis skills) can provide a set of benefits to cover for the very challenges that are introduced. Although it is safe to assume that agile provides answers to most of the waterfall-related challenges, there are two types of challenges in agile projects to be on the lookout for:

1. *Agile-specific challenges*: These are challenges related to the application of agile.
2. *Challenges that linger*: These challenges seem to hang around in the organization and get in the way of delivering value to the customer.

The following sections review those two types of challenges.

Agile-Specific Challenges

These are the challenges that are related to the application of agile:

- *Attempting to go agile in environments that are not suitable for it*: For example, when the project sponsor is not sufficiently involved (in many places the sponsor is often nowhere to be found) or when the feedback cycles are too long.

- *Attempting to go agile in projects that do not need to be agile*: In general, when a project has clear scope, low change, and no need for early realization of value, there is no need for agile methodologies.
- *Attempting to force agile practices even if they are not suitable to the organizational context and needs*: For example, forcing Scrum practices where there is no practical use for them.
- *Attempting to overdo agile*: Using agile methodologies does not mean, or even imply, that we should be aggressive, irresponsible, or reckless in our actions. Agile projects are not the Wild West, it is not a contest of who has the shortest iterations or who works faster, and it is not about forcing processes and practices that the team or the stakeholders cannot handle. Agile concepts can be best applied in a disciplined and responsible way. It is also important to apply agile concepts consistently across the organization and in a way that is suitable—as opposed to attempting multiple flavors of agile and flip-flopping the minute something does not go exactly as initially planned.
- *Failing to properly follow the agile processes*: For example, the daily Scrum meeting is for the team, and therefore, team members should not view it as a drag. The same goes for the retrospective at the end of the iteration— it is in place so the team can come up with, and apply, meaningful lessons.
- *No proper application*: Overfocusing on agile practices, ceremonies, and artifacts to the extent that it consumes the team's capacity to add value and focus on real customer needs. There is a need to be pragmatic and ensure the *right* way to do things according to organizational constraints, team capacity, and customer needs.

Other Challenges

Some challenges linger in the organization and eventually get in the way of delivering value to the customer:

- *Misconceptions*: Agile does not replace the need to know exactly what has to be done—so if there is no clear direction or understanding of the needs, agile will not make things better. It also does not replace the need to have clear objectives. Agile methodologies can help deliver project success, but as the old saying goes: "If you don't know where you are going, any road will get you there."
- *Unrealistic expectations*: Many people still believe that agile is about working fast, having no documentation, delivering early, and having no planning—and as a result, people expect agile to just deliver success with no effort.

- *Organizational challenges*: As agile requires multiple adjustments in many areas across the organization, it is known to be very unforgiving, and it often highlights organizational issues that may appear unrelated to the application of agile concepts, yet are introduced when attempting to apply agile concepts. Examples of organizational challenges include problems with (or insufficient) financial tracking, time tracking, enterprise resource management and reporting, and escalation lines. In addition, in many environments there is no proper adjustment to governance processes to accommodate the nature of agile projects.
- *Resource allocation*: The cyclical nature of the work in agile projects requires resources to be involved repeatedly throughout the project unlike the very structured and phased involvement of specific roles throughout waterfall projects.
- *Challenges with the team's performance*: Many teams struggle to handle the constant changes, different needs for communication and reporting, ongoing stakeholder involvement, the need to work in more ambiguous conditions, and the nature of the cyclical work. Many team members still *think waterfall* and they fail to capitalize on the benefits that agile introduces by not keeping up with reporting, not being proactive, and failing to collaborate. Team members may also find it difficult to adjust to the *open door policy* where the customer is regularly involved and there is full visibility to reporting. Even the setup where the team is physically colocated may introduce challenges since some team members may not find this noisy and hectic setup to be conducive to productivity.
- *Stakeholders*: Many stakeholders struggle with the agile cycles, practices, ambiguity, the need for ongoing feedback, and the changes that agile requires in regard to governance. Stakeholders often find it challenging to adjust to the pace, the cycles of involvement, or the close working relations with the team; they often fail to provide meaningful feedback when required in a timely manner and sometimes struggle in the light of the slightest problem that the team faces—panicking and demanding answers and escalations to issues that do not warrant that kind of attention. Finally, stakeholders sometimes struggle with the need to be involved on an ongoing basis and from a customer's perspective—to accept releases, perform user acceptance testing, provide feedback, and actually use the interim product increment.
- *The extent*: Agile is an organizational change and it requires adjustments in many areas across the organization. Agile is not just a change in the organization's approach to IT or a change in the software development life cycle; many organizations fail to realize this—thinking that things will simply sort themselves out on their own. It is also common to see organizations making partial adjustments to processes, keeping waterfall

project management offices (PMOs) in place, failing to bring in sufficient training and coaching, failing to consider organizational readiness
for agile, or allowing external consultants to *experiment* and provide advice that is disconnected from the organizational context.

Business analysis skills, or a BA, will not simply solve any of the previously listed
challenges, but having team members with business analysis skills can help partially overcome many of them. BAs have an eye for waste and a sense for efficiencies. They complement team members' work and engage with stakeholders.
BAs also map and improve processes, help resolve issues, unclog bottlenecks,
and cover for gaps and disconnects. BAs also bring to the table excellent communication and facilitation skills, along with the ability to see details throughout the big picture. With that said, BAs are not magicians, but when there are
issues, BAs can help minimize their impact, escalate to the right level of support,
or help find a solution.

The Business Analysis Advantage

The expectations of team member's business analysis skills go beyond being
strong communicators and the ability to work beyond role definitions and
across organizational boundaries. In agile environments, team members need
to communicate and share information in real time and on an ongoing basis
as they gain understanding of the elements of work they need to perform in
real time. The incremental understanding and sharing of the changing needs
becomes more formal as part of the backlog maintenance process. This is different than what takes place in waterfall environments where the business analysis
team first tries to a gain full understanding of every aspect of the work, the
needs, and the solution before sharing any of their findings with the development team, the stakeholders, and subsequently with the testing and the quality
assurance teams.

Both agile and waterfall allow the requirements to be presented in a variety
of ways that suit the needs of the team members and stakeholders. They can
be in the form of narratives, diagrams, process flows, use case (scenarios and
diagrams), detailed designs, or wireframes. While in waterfall environments, it
is expected that a BA has the skills to utilize any one or more of these tools; in
agile environments, where the demonstration and application of business analysis skills are shared among the team, it is expected that essentially any team
member would have the knowledge and skills to utilize these tools. The knowledge and ability to apply these skills may vary quite significantly from one team
member to another and some team members may lack most of these skills.

If the agile team has an individual who is a dedicated BA, it can partially alleviate some of the need for other team members to demonstrate business analysis skills; allowing them to focus on their technical work and giving the BA a chance to step in where the business analysis skills are required with the right type of tools to utilize. Having the BA being part of the team can also provide help to the team in the final push toward the latter part of each iteration to meet the iteration's goals. The product owner can also benefit from having a BA on the team by utilizing him or her to enhance communication between the team and the product owner and defining the iteration's goals. The fact that the BA helps the product owner does not imply that the product owner should pull back from working closely with the team, but it can definitely improve the quality of communication and collaboration between the team and the product owner. This is an important point to reiterate to product owners: if they retreat and pull back their involvement with the team because they know that there is a BA to cover this gap, the retreat may cancel the benefits that the BA brings to the table since most of the BA's effort will now focus on the widening gap with the product owner, rather than having the BA step in and help the team wherever required.

Proxy Product Owner and What It Means

It is common to find references to the BA being a *proxy product owner*. While we use this term several times within the book, our intent is to have the BA help the product owner and enhance his or her role where needed. Unfortunately, in many cases, people use the term *proxy product owner* in a different way; thinking that a BA can actually replace the need to have the product owner involved. This is a slippery slope because it may open the door for an increasing number of product owners to think that it would be sufficient to have the BA *replace* the need to involve the product owner. The BA cannot fully replace the product owner since the BA does not have the seniority; the full, big-picture view; the context; or the capacity to make decisions that need to be made by a senior and legitimate representative of the business (and hence, addressing the client's needs). The BA is knowledgeable and capable, but only for what is intended for a BA to do—business analysis work. Having the BA acting as a *full* proxy for the product owner is more of a misinterpretation of the term *proxy* since it takes the BA's role too far in covering for the absence of the product owner.

Further, the BA is unlikely to have the relationships and the rapport that the product owner has with key stakeholders (both internal and external). The BA is also less likely to have sufficient knowledge of the true and underlying business needs of the customer. With a lack of context, a lack of understanding of the full range of the customer's needs, and challenges in accessing key stakeholders, if a BA tries to act as a proxy product owner, it is most likely going to delay feedback and key decisions or even compromise the quality of the decisions made.

To recap: the product owner has to be involved in the agile project on an ongoing basis. The BA can help streamline the communication for the product owner—and even partially reduce the negative effects when the product owner is not sufficiently involved as required—but the BA cannot outright replace the product owner. The product owner has to lead the way with stakeholder engagement, prioritization, and benefits management (especially, but not only financial); provide information for user stories; make decisions about backlog maintenance; and provide timely and effective decision making. These things cannot be done effectively by replacing the product owner with a BA, but can definitely be enhanced by having the BA supporting the role of the product owner as required—through full coordination and collaboration between them.

AGILE AND SOCIAL IMPLICATIONS

Many agile practitioners view agile as nothing short of being the perfect setup for the changing trends in society and with the tendencies, reward systems, and conduct of young people—especially millennials. This section will further discuss the applicability of agile to social trends, but first will address a couple of areas of concern.

Dilution

With more people working in agile environments, more of the *bad behaviors* that are enabled by waterfall will find their way to agile projects. It is simple: when agile was less popular, we generally *handpicked* resources to become part of the team which means we managed to work with the resources that had the most *agile friendly* behaviors and skills. It was an exclusive club of high-performing team members who followed agile processes and practices and ensured that any deficiencies and challenges were addressed the best possible way.

In a few years, when most projects will use agile life cycles, the majority of people in the organization will work on agile projects—many of whom will not be suitable for the agile setup. This does not imply that these people do not have skills or experience, but rather that they may not have the right skills to collaborate, communicate, constantly engage with others, and maintain a proactive approach throughout. The challenges that it will introduce include inconsistent performance across agile projects within the organization and inconsistencies among agile teams within the same projects.

These inconsistencies will dilute many of the benefits that agile brings; and with a breakdown in communication, disconnects from client needs, delays in decision making, and unclear prioritization—it will lead some projects to a *death by a thousand cuts*.

Challengers

As with anything else that is mainstream, with the emergence of performance issues in agile projects and with an increasing number of agile projects that are struggling to deliver on their promises, new approaches will emerge that will challenge the foundations of agile methodologies. When people struggle, any promise for a new and improved way will lure many practitioners to try it. Similar to what happens in politics, when a candidate strikes a chord with promises for change, he or she usually gets people to believe in that change and pursue it. The challenges to agile may or may not bring a solution to the new and emerging plans of product and software development, but the new ideas will most likely lure talent, focus, and resources away from agile. It is unrealistic to believe that something will last forever.

Forcing Agile

Until the challengers to agile introduce *new and better ways*, or *platinum and super pragmatic* approaches that promise success, we will deal with the attempt to force agile where it does not belong, such as:

- Organizations that cannot handle the pace of agile (governance, decision making)
- Customers that cannot handle the need for constant involvement
- Teams that struggle with the processes
- Projects that should not be agile to begin with (clear objectives, no change, no need for realization of value)
- The most potent form of *forcing agile* is when business leaders think that agile can make up for a lack of decision making and prioritization, undefined objectives, or a lack of capacity management. Agile environments need a nimble governance process (yes, this sounds like an oxymoron), timely and ongoing prioritization, clear objectives, and a clear realization and ongoing management of organizational capacity. No agile project (or any other project) will succeed without these aforementioned ingredients. There must be an obvious mandate and the organization must have the ability to allocate the right resources, in sufficient numbers, at the right time, and for the right duration. In fact, agile will prove itself to be very unforgiving as it will quickly expose failing to do these things.

Defying the Laws of Physics

Playing games and trying to defy the laws of physics by achieving goals without making the right amount of investment can happen once in a blue moon: when there are extenuating circumstances or when the stars are aligned and miracles take place. However, the phrase *you get what you pay for* always applies. Business leaders must come to terms with the reality that we need to invest in success, plan for it, and be accountable for our actions.

Scope, time, cost, and quality (part of the competing demands and the main ways to define project success) are called trade-offs for a reason: the organization and the project team must balance these opposing forces and manage the trade-offs among them. This means that when there is pressure on one of these items, something has to give and the trade-off must be to the same extent. In non-agile projects there is a need to remind stakeholders that the trade-offs among the competing demands act in a similar fashion to pressing on a balloon[1]—when you press on one side, there has to be a reaction of an overall equal size on all other sides. While agile *locks in* three of the four elements (time, cost, and quality), there is a need for careful handling of the impact on scope through an effective and timely prioritization process throughout the project. In fact, this prioritization process also acts as a balloon, as the impact of adding an item to an iteration has to be clearly measured by clearing another item (or more than one) of an appropriate size that is in lieu of the item that was just added. The need for the equal trade-off is due to the fixed capacity that the agile project has since it is assumed that there is no ability to add resources, time, or budget (and compromising quality is obviously unacceptable).

Further, capacity management has to start from the planning stages since the team cannot provide plans, estimates for velocity, or properly balance trade-offs without knowing its capacity (up front, at least for each iteration, but preferably for the duration of the project, as much as possible). While for the most part, agile methodologies call for dedicated teams, it is possible to deliver value even with shared teams. However, since agile projects need resources at very specific points in time within each iteration, if a resource is late reporting to the agile project, it may jeopardize the entire iteration's goal.

Agile teams must know how much capacity they have in order to make commitments and to deliver value, and the senior stakeholders must provide the capacity and the commitment to the team. Gone are the days where the project team *saves the day*: agile does not rely on *cowboys* or *heroes* to demonstrate super-human capabilities and somehow deliver success against all odds. In fact, many waterfall environments used to rely on all sorts of heroism by team members to deliver success. Since the current reality points at even more aggressive attempts to cut costs in order to maximize the benefit-cost ratio, decision makers need to realize that they can pull this rope only so much more before it snaps: there is only so much value that can be created if stakeholders persist in their attempts to shirk responsibility, delay decisions, fail to prioritize, and force teams to commit to more than what they can do. We need to remember that when the trade-offs do not balance, the problems will emerge with quality; and in agile environments—it will be sooner than later.

Decision Making and Generational Gaps

Agile needs timely and effective decision making and we can include prioritization under the umbrella of decision making. Decision making is a term that can be evaluated in many ways, including its timeliness, effectiveness, and even efficiency. The decision needs to be made in a way that does not consume too many resources and too much capacity and it must maximize the value of the decision to impact the project.

We are facing many issues with decision making in organizations that are not only in the way of achieving success in waterfall projects, but they are also in the way of delivering value in agile projects. Further, the trend is that issues and challenges around decision making are only growing and becoming more potent. With predominantly three generations in the workplace, there are a few generalizations we can make about each of the generations that have negative implications on agile projects.

Baby Boomers

For the most part, they are *old school*. They do not know much about agile, they have limited familiarity with technological trends and their decision-making process is rather slow. At this stage in their career they are also less inclined, reluctant, and slower to adjust and speed up the way they do things. While not all Boomers qualify under these generalizations, the challenges we listed here serve as speed bumps in agile environments

Gen-Xers

Members of this generation are not strong with decision making. It has become clear that members of middle management, as well as those who are already in the strategic leadership of organizations, lack in their capacity to make timely decisions. Further, many including Gen-Xers have a fear of making tough decisions. These tendencies can significantly interfere with agile projects and the need for clear and timely decision making.

Millennials

It seems as if agile methodologies were designed especially for millennials and the ways they think and act; and this is good news. Table 10.1 provides an overview of how agile concepts align with the way millennials think and act, along with some areas of concern with the agile setup. A BA can help ease some of the issues that are triggered by adapting agile approaches. However, a BA cannot make up for the lack of an agile *state of mind* and to challenges presented by organizational culture and the individuals within it. This is similar to the notion that following an agile life cycle, in and of itself, does not replace the need to have a vision, prioritization of tasks, and a clear mandate.

While Millennials have an increased focus on nonfinancial rewards, a need for transparency, tangible progress, quick cycles, and clear feedback, they also bring some challenges that include struggles with authority, organizational culture, individual work, and the ability to learn and grow in a way that is a win-win for the individual and the organization. With that said, the concepts, practices, and benefits that agile offers are, for the most part, strongly aligned

Table 10.1 Agile and millennials: the good and the not so good

Areas of Consideration	How Agile Aligns with Millennials	Areas of Concern
Work style	Groupwork and collaboration	Challenges with the interpersonal side of the work
	Startup atmosphere	Despite the dynamic pace, there is a need for strong governance and structure
	Ad hoc and dynamic style	May become too dynamic and unstructured to clearly realize value
	Exposure to various sides of the business and team's voice is heard	May lead to challenges with how the team views leadership; especially when decisions that are not in line with the team's feedback need to be made
Mandate and goal setting	Millennials' strong need for a clear mandate	Team members may push for less structure
Desire for feedback	Short cycles that produce tangible results	Coming up with meaningful lessons on an ongoing basis; potential struggle in applying the lessons
	Strong desire for constant feedback, checkups, and praise	Other team members may not want the same type of feedback as millennials prefer
	Suitable for coaching and mentorship for constant feedback and ongoing improvement	Challenges with negative feedback; need to provide positive affirmation first

Teamwork	Progress and deliverables are assessed on a team level with less focus on individual performance	Challenges with feedback per individual since there may be a desire for team-wide feedback
	Tangible progress and opportunity to be creative and to add meaningful value	Channel the creativity to ensure value creation
Reporting lines	Less structure and no traditional hierarchies within the team	May blur career growth opportunities
Pace	Short attention span is suitable for the quick cycles	Strong need to ensure vision is clear
	Quick action, research, and problem solving	Expectation that change and growth should happen at the same fast pace
Transparency	Strong need for transparency is addressed	At times, prioritization and other decisions may not provide sufficient transparency, which may lead to buy-in issues
Work-life balance	Strong fit since agile provides opportunity for flexibility; sustainable and consistent pace assume normal business hours to keep this balance	Flexibility can backfire by not having certain resources when needed and by challenging priority settings
Boss' involvement	The product owner is regularly involved and works closely with the team; Scrum Master is *servant leader*: close working relationships help motivate, create buy-in, and support the team	Concern about crossing the line of *boss as a friend*
Challenge and motivation	Beyond money	May be hard to satisfy all team members
What is agile	A state of mind and an approach to collaboration and value creation	May be confused with working faster and being unstructured to a reckless extent
		May lead to overemphasis of the methodologies and the tools in an attempt to make up for the lack of the proper state of mind
		Too much focus on the ceremonies may also be interpreted as too much structure and push millennials away

with Millennials' style; and if the challenges are addressed effectively by all involved (organizational leadership and individual team members), agile's benefits will strongly outweigh the challenges—especially in the short term.

THE FUTURE

With more organizations moving toward agile, there will be a need for more mature and consistent processes and tools to support the increasing number of agile projects and to check the overall benefits that agile produces on the organizational level. Despite the push for less structure, shorter cycles, and continuous delivery—the need for processes, consistency, and tracking will grow. This is where a BA can own the area of ensuring tracking, processes, and reporting tools to maximize the benefit realization for the organization. In this context, a BA from within the project may not necessarily be able to fulfill the role of overseeing or even leading the activities around the tracking, processes, and tools described here—so it may call for an enterprise BA. An enterprise BA does not focus on the activities within a project and, therefore, does not focus on providing support to agile team members, but rather, the enterprise BA works *outside* of the projects on the organizational (or portfolio) level and acts as a branch of the PMO—or even of a higher level within the organization. With or without this type of BA, the organization will need to ensure that the tracking, processes, consistencies, and benefit measurements are in place across the project.

Automation

The need for more advanced reporting and tracking is likely going to have an impact on another area—automation (beyond testing automation)—in an effort (and a need) to provide consistent and clear reporting, tracking, and trend analysis to ensure that the organization benefits from agile across projects. With the enhanced tracking capabilities, it will be easier to fine-tune processes, improve resource allocation, and better understand the impact of actions within and across projects. It will also help streamline processes, optimize the duration of the iterations, and leverage lessons learned—not only within projects, but also across different projects in the organization.

Distributed Teams

Automation of processes and reporting will be increasingly useful with the increase in the number of non-colocated agile teams. As agile becomes more mainstream, there will be an increase in the number of distributed agile teams.

While there are already multiple tools that allow agile teams to collaborate remotely, these tools (and others) will need to offer enhanced capabilities to consolidate information and leverage it across projects for the organization to be able to interpret the big picture.

While automated reporting and collaboration tools will continue to provide new ways to improve the performance of distributed agile teams, the tools will not be able to completely reverse the negative impact of not having agile teams colocated. The main challenges of agile distributed teams include the following:

- *One-on-one communication*: When teams are not collocated, it is much more difficult to develop relationships and collaborate effectively. The most effective way to communicate is by having two people standing in front of a whiteboard—and virtual environments cannot fully recreate this experience.
- *Building a cohesive team*: Some team members may not be at the same location with the product owner, which may lead to challenges in understanding mandates, priorities, or decisions. The distance may also prevent team members from working effectively together and it becomes more difficult to accommodate different personalities and working styles within the team. Communication, context, and logic may also suffer due to the time difference.
- *Flow of information*: Limited communication channels and longer turnaround times slow down the team's performance and the ability to address issues. Despite the need for more communication in distributed teams, people tend to understand each other less and gradually reduce the amount of communication with the other location(s).
- *Performance*: The communication and team cohesiveness challenges are likely to lead to slowdowns, velocity issues, and struggles in the logic of the work performed. It also becomes harder to ensure that people are on the same page and voice and validate assumptions. Even when the team members work well together, there might be challenges related to access to the users, as well as with knowledge sharing and knowledge transfer—challenges that become progressively more acute with distributed teams.
- *May lead to an us versus them atmosphere*: While collaboration still takes place, people are naturally drawn to develop relationships and work more effectively with those who are physically around them as opposed to those who are in a different location. Over time, team members start to get frustrated with the performance, processes, and styles of those who are not colocated, and this mindset can lead to blame and compromise the team's performance.

- *Feedback and collaboration*: Since people do not see how and what the team members in the other location(s) do, or who works on staff, people become less collaborative. They do not *buy* their remote colleagues' answers, feedback, questions, or timelines to perform work and, in turn, grow apart. People naturally tend to misinterpret each other's information and attribute anything that does not go their way to cultural differences. Therefore, the team members in the different locations grow further apart and continuously reduce the amount and quality of their communication.

Appointing a BA (or possibly a BA at each of the locations) can ease many of these challenges since the BAs can provide support and champion the communication areas that help reduce the negative impacts previously described. It is also strongly recommended that the BAs at each location not only know each other, but also know the team members in the other location(s). Depending on the size of the team and on the project's needs, it may be sufficient to have only one BA—but he or she still needs to be familiar with the team members at all of the distributed locations.

ORGANIZATIONAL AGILITY

Since agile introduces significant impact on a wide variety of areas across the organization, agile adaptations must be viewed as organizational change. This means that there is a need for more *organizational agility* to better accommodate the agile projects and maximize the benefits that these projects produce. Adjustments need to be made in a variety of areas that not only will make things better for the individuals who work for the organization, but will also introduce some organizational challenges: there is a need to strike a balance between the need to allow the team to self manage and the reporting structure. Agile environments need clear reporting lines, governance structure, and a hierarchy, while allowing for flexibility and growth. There is also a need to ensure that while the organization pursues value as part of a strategic plan and that the projects align with the bigger picture, there is room for the team to explore, experiment within the context of creating value, and do so without veering off target.

The pace of change is going to continue to grow, therefore, organizations will need to be fast on their feet in reinventing their purpose and evolving as their needs change, instead of providing static missions that are part of long cycles. With the faster pace of change, projects and products will need to be more evolutionary, more integrated with each other, and flow continuously—as opposed to having a number of projects where each one is managed as a silo. The faster pace and the need to integrate projects across the organization are reasons why the company can benefit from BAs—both within the projects as well as on the enterprise level—to ensure consistent processes and practices within projects and consistent evaluation, tracking, and flow across projects. With that said, the product owner and the project team will need to further enhance their collaboration and the mechanisms used to incorporate feedback into the decision making and prioritization processes.

With agile projects requiring engagement, buy-in, conformity, and capacity, there is no need to look at research to know that many employees are not actively engaged, many believe they can find a better job (and are actively searching), and most of them are overwhelmed with the amount of work they currently have. These areas must improve before organizations will be able to increase the number of agile projects and realize organizational success in delivering value through these projects. According to the 11th Annual State of Agile Report,[2] as seen in Figure 10.1, many organizations experience challenges in adopting agile that are related to organizational issues. In fact, three of the four most common challenges are related to organizational factors, such as company philosophy and culture, lack of management support, and resistance to change. These types of organizational challenges will continue to stand in the way of delivering value through agile projects if not properly addressed.

Interpersonal Skills

Whether it is the product owner, Scrum Master, PM, BA, or any member of the team, there is a need to increase the overall focus on soft skills. These skills are nothing short of critical to the success of agile projects and the production of value and benefits for the organization and the client. With an increasing

*Respondents were able to make multiple selections.

Figure 10.1   Challenges in adopting and scaling agile

number of mundane and low-risk activities (and even projects) becoming auto-mated, there will be added value to those who can demonstrate strong interper-sonal skills in performing more risky activities and projects. The application of the interpersonal skills will be different, based on the role that each individual plays in the agile environment, with some degree of generalization. Product owners and members of senior management need to have the ability to develop and share a clear vision of success and to communicate this mandate. PMs and Scrum Masters will need to further enhance their conflict management and res-olution skills. While teams need to be self sufficient, there are conflicts that cannot be resolved within the team and it is the person who leads the team who needs to assess, manage, and support it in bringing the conflict to a resolution that is in context of the situation and the stakes.

Everyone involved with the agile team needs to work on enhancing their ac-tive listening and ability to empathize as part of the process of handling change, understanding each other, building consensus, and making decisions. We need to keep in mind that although decisions about the product need to be made by the product owner, the team needs to make many decisions about recom-mendations, courses of action, impact assessment, escalation, and performance. Active listening, the ability to fully understand someone else's point of view, as well as motivating and influencing skills become critical in these situations. Another important skill is the ability to network and navigate one's way through organizational boundaries. Everyone involved in agile projects needs to develop skills to sense underlying forces and trends, as well as other political issues, in order to ensure that we engage the right people, in the right context, with the right timing, and to the right extent.

Needless to say, the BA should possess as many soft and interpersonal skills as possible to provide support to any individual role around the agile team, as well as to cover for and address challenges that may appear throughout any type of interaction within or around the agile project. While it may not be possible or realistic to expect a BA to have all of these skills, identifying and realizing the impact of organizational processes and environmental factors can help find the best-fit BA for the needs of the project and the organization.

Discipline

Many organizations *unleash* agile aggressively, thinking that running forward with no rules is the right way. The result is project failures, failing to deliver the intended value through agile, and at times—even blaming agile methods for the failure. Subsequently, in an attempt to improve things, these organiza-tions end up blaming the messengers, firing people, and desperately looking for reinforcement by bringing in a mix of consultants with different flavors of agile and widely varying approaches to fix the problems. This, in turn, brings more problems, gaps, misconceptions, and eventually more failed attempts to go agile. In the meantime, money is being wasted, people lose their jobs, frustration grows, morale hits new lows, and projects continue to fail—exacerbating the problem and deepening the snowballing crisis.

The answer to this scenario is to slow down, perform a proper readiness assessment, develop criteria to determine if a project is suitable for agile, and proceed with a disciplined approach to adopting agile. Critical success factors in these cases include consistent executive support, building internal agile coaching capabilities, applying consistent processes and practices, and properly utilizing a responsible external agile trainer or consultant.

Disciplined agile delivery (DAD)[3] principles call for simplified process decisions around incremental and iterative delivery, moving beyond Scrum. DAD is a "hybrid agile approach to enterprise IT solution delivery that provides a solid foundation from which to scale."[4] While this approach also does not name a distinct role for the BA, it calls for a hybrid application of agile methodologies with enterprise awareness and a focus on learning (to name some of its key aspects). These can imply a need for a BA to improve in these areas. DAD's approach to project roles also implies that there is room for a BA under one of the secondary roles it names (could be integrator, specialist, or domain expert).

Social Trends

Looking toward the future cannot be done without addressing some specific social trends that are taking place in front of our eyes. And while agile appears to be *perfect* and in line with these trends, there is a chance that these trends will overwhelm agile projects, and subsequently, organizations. The discussion in this subsection is an interpretation of the underlying forces that are a part of emerging social trends and many of the potential consequences described here may take time to materialize.

Hierarchy

One emerging organizational trend points at less layers of reporting, along with attempts to reduce the amount of structure and hierarchy in organizations. Agile aligns with these trends by offering a *flat* project organization with team members being virtually equal in title (with the exception of a Scrum Master to enable the team to achieve its goals and a product owner who is in charge and accountable). While there may be team members who challenge even this type of structure, for the most part it allows the team to thrive.

However, when breaking an organizational hierarchy, it may introduce different types of challenges (e.g., accountability, reporting, and escalation issues) and if some individuals do not perform their role properly (e.g., product owner who is not sufficiently involved), it may break down the entire construct. A BA can help *cover* some of the potential deficiencies by complementing certain roles that are in need. For example, as we discussed earlier in this book, the BA can streamline communication and provide other support to the product owner.

Openness

Open communication and collaboration are not only critical to agile, but they also are signatures of younger team members. Although the trend is very beneficial for agile success and can thrive in agile environments, there have been notorious problems in regard to communication. The abundance of social media and communication channels has not made us more effective or efficient in our communication, and the all-time struggle of distinguishing between formal and informal communication may get even more challenging as communication lines, formats, hierarchies, and organizational structures become blurry. A BA cannot undo this damage, but he or she can help control and streamline communication and facilitate the handling of formal versus informal communication.

PMOs

With the breakdown of organizational boundaries, hierarchies, and structures, the PMO may be one of the most obvious victims. For the most part, PMOs have not proven to add meaningful value to organizations. The reporting lines usually have the PMO reporting to higher levels in the organization, but this speeds up the process of the PMO losing touch with projects since the PMO becomes increasingly focused on addressing their superiors' needs as opposed to the project's needs. Many PMOs struggle between *taxing* the projects and adding more red tape versus adding value and helping projects; and the PMOs policies and processes are often disconnected from the actual needs of the projects.

Perhaps the most potent issue surrounding PMOs is that their structure and focus areas (that have already been under pressure for not addressing project needs) increasingly struggle to accommodate agile needs and align with the cyclical nature and fast pace of agile projects. As a result, the PMOs struggle to actually produce value for agile projects—with many PMs and practitioners questioning the very need of having a PMO. Their thinking is that with empowered, self-sufficient, and self-directed teams, there is no need for another governing body to oversee the project's performance and boss people around.

However, dissolving PMOs may be premature because—if done properly—PMOs can help provide an important layer of governance, capacity management, oversight, support, guidance, and application of portfolio management functions—specifically prioritization within, but especially across projects. Further, with no PMOs, knowledge management, information repositories, and application of best practices may be lost and projects may start to pay a price as a result. A BA can provide a partial relief in areas that are no longer addressed by the PMO. In fact, the declining role of a PMO may be replaced by an increasing role of a business analysis office (BAO), which would be a center of excellence for BAs and for those who practice business analysis in agile projects.

With no need to oversee the *traditional* areas that PMOs did, the new body (the BAO) will start to grow in its formality, processes, and structure—until it practically replaces the PMO. While the BAO will not have the authority or clout that PMOs used to have, it is a more suitable structure to help with collaboration, sharing ideas, support, suggestions, application of best practices, and even a place to seek emotional support for ongoing project struggles. Unlike the PMO that worked almost exclusively with the PMs, the BAO will provide support to essentially all team members.

Inertia

As always, there is a *but*. Changes to organizational structures and hierarchies do not and will not go smoothly. There will be resistance (from all generations, including millennials); and with unclear reporting lines, performance challenges will start to emerge. There will be an increase in friction within and across teams, conflicts (that will be harder to facilitate and resolve), confusion, unclear mandates, broken communication lines, conflicting priorities, and unclear escalation criteria and rules of engagement. For a while, most projects will continue to perform well, but the degree of success will start to erode. Further, for the most part, the projects' success will be owed to inertia—the ability to move forward despite the new realities—as opposed to strong underlying performance. Projects will still be surviving, but it will be increasingly because of the heroics of team members, and less because of a sustainable, stable, and consistent approach to project management.

With organizational structures and norms imploding, agile methodologies will be challenged. While agile is showing a multitude of benefits under the current social and organizational structures, it is hard to see how we continue to improve our projects' rate of success when the entire system is being challenged. At that point, agile alone will not be enough, the inertia will dissipate and the new messy reality will catch up with us.

Business analysis is an area that is aligned with emerging social trends and is beneficial when it comes to dealing with the impact of them. Unfortunately, business analysis alone will not be able to hold things together. Improving business analysis skills among team members and introducing more BAs into projects will help soften and delay the downsides of future social and organizational changes. By delaying the impact, it will buy teams and organizations more time to find new ways to deal with these new realities.

THE FUTURE AND THE BA

With some adjustments that enable more organizational agility and a more disciplined approach to agile, there is a better chance for more projects to succeed in their agile form even as the number of agile attempts continues to grow. Further, with the growth of agile, organizations may face challenges in areas that include the ability to track and control agile projects on an organizational level and the ability to properly manage stakeholder expectations. While managing expectations is typically the job of the project lead, PM, or Scrum Master (with support from the product owner), the BA can step in to help with this important task based on the strengths that the BA brings to the table. This can include the ability to understand stakeholders' needs, the ability to empathize and relate to stakeholders in search of a better way to provide clarifications, and the ability to provide access to subject matter experts (SMEs) and resources to provide more information and context. While the BA is not likely to lead the efforts around managing stakeholder expectations, in some cases the BA may do so depending on the context and specific stakeholder's needs.

In cases where the agile team utilizes a distinct BA, the BA must be in sync with the PM (or the Scrum Master) since these two roles must work in harmony and with a level of coordination that makes them appear as one entity[5] to others in the project. While this is not a call to integrate the roles of the PM (or Scrum Master) with that of a BA in one person (unless it is sufficient for smaller projects), having the PM-BA duo work seamlessly together as one entity on the larger, more complex initiatives has generated very favorable results with an overall focus on what agile projects focus on—delivery of value and risk reduction. Incorporating the role of the BA in agile projects is part of a bigger need to be pragmatic about agile and to apply it based on the organizational and customer needs—as opposed to trying to apply agile projects uniformly with limited regard to the project's context. If done correctly, having a BA in the agile project can redefine the BA's role into a center of value, ensuring agile project success.

Going back to the question of whether a project needs a distinct BA or whether it is about ensuring that team members demonstrate business analysis skills, there are a few areas of responsibility that fall under the *umbrella* of the BA, led by the following:

- Identify user stories
- Facilitate and lead workshops
- Conduct interviews and perform other information-gathering and requirements-elicitation activities
- Specify minimum viable functionality for stories
- Analyze stories and ensure appropriate SMEs get involved
- Utilize the right tools and models for the analysis process
- Help define acceptance criteria
- Ensure that all stories slated for an iteration are clear, understood, and estimated
- Ensure stories are included in the backlog, in priority order, and with acceptance criteria
- Work with team members and stakeholders to handle changing needs, changes to stories, and backlog maintenance
- Ensure detailed requirements are communicated to the team and that communication channels are open for planning and clarifications
- Document stories and ensure all relevant information is stored and managed properly for the development effort, as well as for future needs and maintenance considerations
- Lead the identification, application, and review of lessons and areas of improvement
- Ensure testers are involved at the point of defining acceptance criteria and subsequently help testers write test cases, automate, and perform the testing as required
- Help escalate issues and identify the right SMEs, stakeholders, and team members to get involved
- Oversee tracking and reporting to ensure progress, tasks, velocity, and status are properly reported to reflect the true project state on a timely basis
- Ensure decisions and actions are documented and that everyone involved had a chance to contribute to both the demo and the retrospective

In addition to the items listed here, there are additional areas where the BA can add value to agile environments, including helping to handle and manage ambiguity and addressing misunderstandings and communication breakdowns that are introduced as a result of agile processes. Also, as part of current social trends, agile methodologies call for *open doors*—having the technical team members work regularly with stakeholders. This adjustment cannot happen in one day, and it requires focus and effort. In the past, developers sat behind closed doors away from the customer—there was no transparency and the technical team's deadlines were not primarily business driven. However, now there is an increase in the belief of the value of relationships—including having all roles within the team working with the customer. In many cases, a BA can help coach team members, contain stakeholders' anxiety, and streamline the communication between all parties involved—while at the same time, managing expectations and even softening the style and the extent of messages and areas of impact.

Another adjustment takes place with the expectation that technical team members now need to plan, estimate, communicate, commit to results, listen to the customer, talk to stakeholders, and provide quick turnarounds for inquiries, issues, and deliverables. Many technical team members still struggle when dealing with any one or more of these areas and the BA can supplement those skills by serving as a check and balance mechanism that helps put things in context, research for information, and adjust expectations as required.

Another area where a BA can provide value to the organization is continuity. Since it is common practice to move from one organization to another, it is expected that many software developers will move to a different organization and not stay at the same job for more than two or three years. This means that organizations lose a lot of knowledge and context every time someone leaves. With a BA in place, it can become easier for organizations to manage knowledge and ensure that information is properly maintained and documented for continuity purposes. This can reduce the negative impact that organizations may face with the loss of a team member. Another aspect of continuity that the BA can address takes place in the event that a team member leaves the organization mid-project. With evolutionary design and an incremental approach to product development, a loss of a resource may spell trouble to a project that relied on this individual's knowledge, and more so—context. Requirements, needs, and product features evolve and change as the project progresses, and unfortunately, changes in the team may break down the flow and the overall understanding of the customer's needs. With a BA in the project, losing team members may still be painful, but the knowledge, information, and context that the BA has are going to reduce the negative impact of the loss.

The Future Is Here

As agile becomes more mainstream and as more people in the organization engage in agile projects, more people are likely to bring their tendencies and bad (waterfall) behaviors with them—and along with it, they fail the system they are a part of. Now that agile is addressing many of the challenges associated with waterfall, we are beginning to find out that it was not waterfall's fault, but rather it was that waterfall simply enabled certain behaviors and allowed some less-than-ideal practices to *hide* within it. Some of these practices include the following:

- Procrastination
- A lack of prioritization and with it a lack of accountability
- No capacity planning—scope was so big and the planning horizon so long that it was not a big deal to take on something else
- No proper documentation
- No management of assumptions (and therefore gaps in the risk planning)
- Big plans, slow decision cycles, and slow response to change increased the risk of building the wrong thing or building defects into the product
- Temptations to multitask (thinking it was a good thing)
- Working in silos and failing to involve testers early enough
- Sponsors that were nowhere to be found
- Weak decision-making processes
- Strong need to remove ambiguity by building long-term, baseless plans

These practices were embedded in waterfall projects since there was no direct impact that was felt in the short term—and as a result, many of the decisions that were made were not the right ones and did not best serve the needs of the customer:

- *No decision?* When things were not prioritized, we were always able to kick the can down the road: no decision?—no problem—we will figure it out later
- *Planning?* We will finalize all details of the requirements and the plans now, so we can secure resources and funding
- *Don't feel like working on something today?* No problem—the deadline is months away and we will make up for it later
- *Change?* Let's assess the impact in real time, while we do other things; start working on it before we get an approval, and if it is approved—it becomes the most important thing to work on
- *Changing conditions?* We will work faster, smarter, and better to accommodate
- *Trending toward failure?* No need to tell anyone and we will remove items as we see fit; we can also suggest that we deliver the product later with some broken functionality

It is most likely that none of these *conversations* actually happened, but the attitude toward delivering value for the customer may have become *loose* in many organizations—compromising the value delivered to the customer and, at times, failing to align project actions with customer needs.

Since agile is becoming more mainstream, many of these less-than-ideal practices may find their way into agile projects with people trying to challenge the new agile *status quo*—either in a search for a better approach to product and software development or in an attempt to simply challenge the mainstream way of doing things. After all, agile appears to have been developed by micromanagers who had poorly performing teams that could not be trusted working for them. It appears that way since agile has so many built-in mechanisms that check on a team's progress on an ongoing basis, not allowing any team member to move too far without checking and reporting about it. The result is an approach where, although they are excessively micromanaged, people seem to like it and the managers who micromanage do not admit that it is micromanagement—they call it sprints, iterations, stand-up meetings, task boards, and other cool names. The result is not quite so bad, but with time, more people will realize how to cut corners and take shortcuts that bypass the strong controls and processes that agile has in place.

This takes us back to the need for organizational agility and a shift in culture that is done through organizational change. Calling things by names, announcing that we are now agile, investing a lot in new job titles and facilities, and even changing corporate colors will not fix broken things in the organization or make them better. When the mandate is not clear, the stakeholders are not set on the requirements so they keep changing them, the product owner provides no leadership or direction, and the team does not grow and improve—it's like putting lipstick on a pig; it may look nicer, but it's still a pig.

Processes and Exceptions

In order to enable creativity, a dynamic environment, and the ability for quick turnarounds, agile requires a lot of behind-the-scenes structure, governance, and leadership; and requires the team to perform consistently and follow agile processes. Many of the problems we suffer from in waterfall environments are a result of people simply not following processes. The reasons behind failing to follow processes vary dramatically from one environment to another, but whatever the reasons are, the results are the same: problems, defects, conflicts, and failures. While sometimes there are more *legitimate* (yet not acceptable) reasons for not following processes (e.g., the process is broken or full of redundancies), at other times, processes are not followed due to less *legitimate* reasons, such as a lack of knowledge about the process, a lack of understanding of the context, not realizing the impact of not following the process, or simply laziness. Sometimes people do not follow a process because they think it does not apply to them, or because they feel special, entitled, or that they are an exception to the rule. Since agile success is based on trust and consistency, if people do not follow processes and cannot be trusted, we cannot sustain performance over time. This should not be an acceptable standard to hold in agile environments, nor in any other type of social construct.

Disruptive Forces

Agile will continue to grow, but as it becomes more mainstream (waterfall will never, nor should it ever, completely go away), people will find shortcuts and ways to bypass the way it should be done. Eventually the tide will turn, and people will say that there is a need to change the way we manage software and product development, and that we should put more strict processes and checks and balances in place that will not allow people to break the rules. People will start getting offended if the processes they put in place are not followed, and they will incorporate penalties against those who break the rules. At that point, new approaches will emerge by offering more controls with the appearance of freedom and choice, and they will have a new name. Another potential trend might involve a pendulum swing to reverse the trend of less documentation, less planning, less estimating, and less structure. At some point, when agile processes are increasingly bypassed, projects will start to default at higher rates, and the call will be for a change in direction—and for more red tape, structure, and restrictions to address those deficiencies. But until then, we will have business analysis skills, or even a BA, to help bridge the gaps, and many practitioners will get down from the high horse they were on when they made the claim that agile had no need for a BA.

We can already see signs that, in some places, agile is faltering—and not because agile is not good, but because organizations are going agile across the board without checking whether it is suitable. They go agile for the sake of going agile—in and of itself—and the agile ceremonies take place, but with little to no substance. In many places the retrospective process is meaningless, with no steps for improvement, and no tracking or controls on the attempts to change things. Many teams fail to have their retrospectives offer constructive discussion; they settle on sweeping statements, jargon, and high-level vagueness instead of specific action items. Further, due to leadership and team performance issues, many planning and estimating games are *rigged*, or driven by collusion and alliances. Scrum masters do not properly lead the team and provide little protection from outside interference. Product owners do not understand their roles, and when asked about requirements or prioritization, they say, "I don't know what we need, I just hope to get it out of the gate in a couple of weeks." But—it is not all doom and gloom: what is described here happens in too many places, but not in the majority of organizations. It is not too late to change course, improve processes, and do it the right way. In some cases, a BA can make the difference in freeing up just enough capacity—for team members, for the Scrum Master, and for the product owner—to change the momentum and the direction of the project (and the agile practices).

The *Right* Way

Doing agile the *right* way—basically, a disciplined way that is suitable for the organization, its needs, objectives, context, teams, capacity, capabilities, and customers—can provide numerous benefits. For example, it can reduce the risk of building the wrong thing; reduce the risk of building a product that is riddled with defects; increase productivity; reduce overall time to delivery; help realize value earlier; improve quality; respond effectively to change; reduce scope efficiently, based on actual needs and priorities; address true customer needs; improve team morale (and retention); potentially shorten the schedule and reduce overall costs; and improve customer satisfaction.

Clearly, agile does not offer guarantees in delivering success, but the aforementioned list is made of a great set of benefits, and even if any one of these benefits is not achieved in full, the majority of organizations doing agile projects can show improvements in their performance and customer satisfaction. While there are many ways that teams can deliver value, the full potential that agile can offer is often being held back—with senior leadership and management levels of the organization being the reason. As Joseph Juran said, "It is most important that top management be quality-minded. In the absence of sincere manifestation of interest at the top, little will happen below."[6] Many organizational leaders believe that just by allowing the mandate to go agile, things will change—and for the better. As was stated previously, following ceremonies, building special facilities, and calling these things by cool names will not make a difference on their own.

So, what is the fix? This book provides answers to address or overcome virtually any challenge that agile introduces. Some of the answers involve application of business analysis skills, others call for the installation of a distinct BA role at the project or the enterprise level, and some of the answers are not directly related to business analysis, yet a BA can help apply and enhance the recommendations. On a high level, any organization that embarks on adopting agile methods or attempts to manage a project following agile principles should ensure that the right approach is chosen, the organization is ready for agile, the organizational change management aspects of agile are taken into account and addressed, and above all, the change starts at the top—with the leadership of the organization and the governance processes—otherwise agile will fail to deliver on the promise and expectations it can bring.

The Stealth Approach[7] is a good way to introduce agile processes through the back door if facing resistance when implementing new development process changes within an organization. Sometimes big changes need to be packaged into small pieces to make them appear as minor adjustments to the current process. This approach includes specific tips on introducing new development methods to the team and the organization without introducing new and potentially scary (agile) terms.

So what do we need in order to make success happen? Start by setting clear goals. Then comes a series of items that can be useful in any type of environment: accountability, collaboration, decision making, and prioritization. We continue with establishing mechanisms for effective dialogue between the customer and the technical team; ensuring that the technical team provides all necessary technical feedback, but not allowing the technical team to make prioritization and product-related decisions on their own. In addition, there is a need for training so team members know how to conduct themselves in agile environments and improve their interpersonal skills. Training can also help the BA and other stakeholders know their role, their contribution, and the value they need to produce. Decision making and prioritization must involve cost-benefit-risk considerations and resource allocation should strive for dedicated teams so there are less resource-allocation games that are part of the perpetual attempt to over-squeeze the matrix-organization lemon. When it is not possible to have dedicated teams, there must be cross-project collaboration for the shared resources in order to ensure that the right resources end up reporting to the project on time and for the right duration. Cross-project collaboration must be governed by senior levels in the organization and supported by BAs.

The BA and DevOps

DevOps[8] (short for development and operations) is a mix of practices and tools in an attempt to enhance the organization's ability to deliver applications and services at a high speed. The main principles behind DevOps support evolving and improving products at a faster pace. Agile and DevOps have lived side by side with some people viewing DevOps as a part of agile, while others see it as a better form of agile—or even as a better way to deliver software. As agile keeps growing in popularity, so does the application of DevOps.

The main focus of DevOps is to bridge the gap between development and operations, and to streamline and reduce the notorious painful handoffs so that things such as deployability, scalability, and support are all addressed proactively and not after the fact. But, as expected, there are some challenges that DevOps introduces in relation to the separation between the development and DevOps teams—creating yet another silo that hurts collaboration. Most agile teams do not include the operations, support, or infrastructure specialists, and there are still challenges with coordination and collaboration within the team. This means that when there are talks or attempts to go DevOps, there is going to be an even stronger need for business analysis skills, and—considering all of the challenges that are introduced by DevOps—for a BA.

The BA and the Retrospective

At the end of each iteration, the team needs to come up with ideas for improvement by having a discussion where its actions are documented and applied primarily by the BA. In the upcoming iteration, the BA will track the application of the lessons and the implementation of the ideas for improvement; and in the next retrospective, the BA will provide a report with the findings on the lessons' application, the benefits realized, and the side effects that were created. Although the BA does not make the decision as to which lessons to pursue and how (ultimately it is the product owner's decision), the BA can contribute ideas to the discussion, as well as context for the team to ensure they pick the right battles. It is important to have only so many ideas for lessons be applied after each iteration—otherwise, significant learning opportunities will be lost.

Of all the questions that can and should be asked at the end of an iteration, it is important that the BA helps the team maintain focus to ensure value creation and maximization. There are a handful of areas that require systematic attention during the retrospective:

1. What went right in this iteration? Why?
2. What went wrong in the iteration? Why?
3. What was a surprise? What ended up being okay despite a lack of planning? Was luck a factor in achieving success? If so, where did luck replace good planning?
4. Did the iteration differ significantly in any way from its goals? In what way? What was the impact of this?
5. Was there anything specific in this iteration that warrants our attention? Processes? Team work? Agile practices? Collaboration? Communication? Stakeholder engagement? Prioritization? Feedback? Capacity issues? Escalations? Defects? Reporting? Turnaround times? Governance? Organizational influences? Cross-project interlocks? Decision making? Customer issues?

Many agile practitioners view the last day of the iteration, as well as the first day of the next iteration, as a waste. These days are full of all sorts of chores and activities that help the planning process, ensure feedback is clear, and facilitate planning and estimating process for the next iteration. These days (last of one and first of the next iteration) are filled with activities that while adding value are not about producing features and making product progress—which are the ways the customer measures progress and value. The retrospective is one of these activities (along with the demo, incorporating feedback, backlog maintenance, accounting for change, planning, estimating, and ensuring everyone is on the same page). In many environments, there is a push to make these two days (last and first) shorter and incorporate more product work into them at the expense of the other activities. The push to *shorten* these days by allowing the production of more features at the expense of these so-called *administrative* chores is dangerous since people need to understand that the activities on these days are very valuable for the team's progress and for the agile process. Shortening these activities may have a negative impact on the team's ability to incorporate feedback, plan for the next iteration, and ultimately, produce value for the customer by building the right thing or by doing it right. It is understandable that the team wants to focus on building product increments and features, and to maximize the time it takes to do so—but it is important to provide support for the product-related progress with sufficient planning and process work. This is the responsible thing to do, and while Scrum provides specific time-boxes for each of the last and first days' activities, the team must determine the appropriate duration for each of these activities based on the customer's and the team's needs and best interests.

THE BA IS NOT THE PRODUCT OWNER: THE BA IS THE BA

Among agile practitioners there are many ideas, misconceptions, and confusion about the relationship between the BA and the product owner. We addressed the area of the relationship between the BA and the product owner throughout this book, and our conclusion was that the BA is not the product owner. While there are many touchpoints between the BA and the product owner and the BA provides valuable support to the product owner, the BA should not act as the product owner. When the product owner is not sufficiently involved in the project, the BA can serve as a valuable liaison between the team and the product owner. The BA can also serve as a translator and the go-to point for team members when they are in need of the product owner's knowledge or feedback. In these cases, the BA can consolidate the message, clarify some points, seek feedback from SMEs, or reiterate feedback previously given. The BA can also interpret certain elements of the feedback previously provided by the product owner or increase the level of urgency in getting the product owner's involvement; but the BA should not make decisions on behalf of, or instead of, the product owner. The role of the product owner is, in a way, the business SME. The product owner is ultimately accountable for the project and for the product's vision. As part of that, the product owner is in charge of requirements prioritization to ensure business decisions are aligned with maximizing customer value. The BA works within the mandate provided by the product owner and is involved in *translating* the business mandate to help the technical team perform the work that will ultimately conform to requirements and be fit for use. To clarify, the BA is not a decision maker, but rather a facilitator, translator, consultant, support person, process and efficiency champion, and trusted colleague.

To recap, beyond the help to the product owner, BAs can benefit agile environments in a variety of ways—led by the following:

- Perform needs assessment
- Help establish a product vision
- Help articulate the business problem
- Lead the requirements process
- Facilitate planning and estimating activities
- Support the creation of user stories
- Ensure that user stories are traced to business value
- Ensure that there are no gaps in requirements
- Map the current and the desired processes
- Facilitate the discussion between the team and the product owner to incorporate technical and other considerations into the prioritization process
- Lead the use of supporting tools and modeling techniques
- Help define acceptance criteria in support of both the delivery team and the testers
- Support the team in process improvement and application of lessons
- Ensure user needs are captured and validated
- Support the team in planning and estimating activities
- Facilitate the clarity of feedback and subsequently the backlog maintenance activities
- Work with the Scrum Master (or the PM) on organizational change
- Support communication, tracking, and reporting within the team and outside of the project as required
- Requirements and solution validation
- Help manage resource needs
- Ensure that there is a clear definition of *done*

One way or another, each of these functions is important in both agile and non-agile projects.

It is not expected for the BA to perform all of these activities on an ongoing basis, but rather when the team identifies its needs. The BA is not introduced in order to pick up the slack, but to complement the team where he or she is needed the most. In many cases, there is no need for an actual BA, but rather for the team to determine which members of the team need to perform what types of business analysis activities in support of the project. If there is an expectation that the BA takes over more than what he or she can handle, the BA will not fulfill the intent behind having a BA, and ironically, the BA will be the bottleneck in the agile project. The mix of areas to look after and the actual role of the BA will be unique depending on the organization; and in addition to the team's role in identifying their need for a BA, it is the BA who needs to define what the role entails.

There is an interesting question to ask: will there be a difference between the BA's day-to-day activities from agile to waterfall projects? The short answer is: yes. The activities will be different and cyclical by nature. However, philosophically, the BA will be there for the team the same way the BA is there for the team in waterfall environments. The difference is that in agile projects, the BA will perform some activities in support of specific agile processes in a similar fashion to how the BA supports waterfall-related processes on a waterfall project. One difference is quite noticeable though: in agile projects the BA supports the process of prioritization and many of the BA's activities are related to it—backlog maintenance, capturing feedback, and facilitating the discussion between the technical team and the product owner, as well as within the team.

The BA will enhance processes in agile projects, as opposed to compensating for team, resource, and organizational deficiencies that are rampant in waterfall environments. The enhancements involve communication to ensure that information flows in the right direction and at the right pace, and increasing the team's ability to maintain focus on value creation by complementing team members' skills in whatever areas are required.

And finally, while BAs provide tremendous value to agile projects—whether there is a BA in the project or not—all team members need to be able to perform business analysis activities. In fact, even if there is a BA within the agile project, team members still need to perform business analysis activities to complement the BA (as well as each other's roles) since the BA cannot possibly cover all of the business analysis needs that will emerge in agile projects. While there are challenges with the need for technical team members to also be BAs, it is important to identify team members' skills and experience, and their applicability to which types and to what extent they need to also perform business analysis activities. While agile is becoming more and more mainstream, so is the need for business analysis skills in the project—by the BA and by virtually all team members as well.

ENDNOTES

1. Adopted from Ori Schibi. (2013). *Managing Stakeholder Expectations for Project Success.* Chapter 2. J. Ross Publishing; Plantation, FL.
2. VersionOne. The 11th Annual State of Agile Report. (2017). http://www.agile247.pl/wp-content/uploads/2017/04/versionone-11th-annual-state-of-agile-report.pdf.
3. Adopted from Scott Ambler and Mark Lines. (2012). *Disciplined Agile Delivery: A Practitioner's Guide to Agile Software Delivery in the Enterprise.* IBM Press.
4. Philippe Kruchten and Paul Gorans. (2014). "A Guide to Critical Success Factors in Agile Delivery." (Report). IBM Center for the Business of Government. p. 14.
5. Adopted from Lee Schibi. (2015). *Effective PM and BA Role Collaboration.* Chapter 1. J. Ross Publishing; Plantation, FL.
6. http://sixsigmastudyguide.com/juran/.
7. http://www.agilepm.com/stealth-methodology-adoption-book.
8. Concepts in this DevOps discussion are adopted from https://www.infoq.com/articles/merging-devops-agile.

INDEX

Note: Page numbers followed by "f" indicate figures and those followed by "t" indicate tables.

Data entry clerks, 130
Data flow diagrams, 139
Day plans, 179, 254
Decisions/decision making, 91, 166,
  279
  cost-benefit-risk considerations,
  297
  generational gaps and, 279–282
  issues with, 279
Defects (bugs), 17–18
  backlog, 202
  dealing with, 232–235
  eliminating/reducing, 18
  potential causes, 18
Definition of done (DoD), 215–216,
  228–229, 246, 248
Delays, 16–17
  eliminating/reducing, 16–17
  potential causes, 16
Demo, 203–210
  challenges and issues, 207–208
  essence of, 208
  feedback, 208–210
  solution, 207
  system, 206, 206f
  team, 205–206
  testers participation in, 246–247
Deterministic approaches, 1, 2, 45,
  46f, 174
DevOps, 29, 31, 298
Diagrams, use cases, 133, 135f
Digital prototypes, 138
Dilution, 3, 68–72, 276–277
Discipline, 286–287
Disciplined agile delivery (DAD),
  287
Disruptive forces, 295–296
Distinct days, 253–254
Distributed teams, 282–284
  challenges of, 283–284
  documentation and, 151

Documentation, 30, 36–37, 145–159.
  See also Reporting
  balance between too much and
  too little, 36–37, 150
  business analyst (BA) and, 145–
  146
  compared with waterfall/
  traditional, 149
  compliance, 150–151
  description, 146–147
  distributed teams and, 151
  excessive, 36–37
  formality levels, 151
  groups/categories, 148, 149t
  overview, 146
  principles, 147–148
  process, 148–149, 150f
  technologies used, 151
  wrong reasons, 146
Documentation monsters, 36
DoD. *See* Definition of done (DoD)
Done in agile projects, 9, 228–229,
  229f
Due diligence process, 48

Earned value management (EVM),
  158
Empirical approach, 1, 47–48
End-user requirements, 94–95
Enterprise analysis, 166
Entity relationship diagrams, 141
Epics and user stories, 126–127
Estimation, 21–22, 55
  complexity of items, 183–187
  ideal time, 183–184
  story points, 184–187
  testers and, 246
  velocity, 188–193
EVM. *See* Earned value management
  (EVM)
Evolutionary design, 55